Circular Economy and Sustainability

This book series aims at exploring the rising field of Circular Economy (CE) which is rapidly gaining interest and merit from scholars, decision makers and practitioners as the global economic model to decouple economic growth and development from the consumption of finite natural resources. This field suggests that global sustainability can be achieved by adopting a set of CE principles and strategies such as design out waste, systems thinking, adoption of nature-based approaches, shift to renewable energy and materials, reclaim, retain, and restore the health of ecosystems, return recovered biological resources to the biosphere, remanufacture products or components, among others.

However, the increasing complexity of sustainability challenges has made traditional engineering, business models, economics and existing social approaches unable to successfully adopt such principles and strategies. In fact, the CE field is often viewed as a simple evolution of the concept of sustainability or as a revisiting of an old discussion on recycling and reuse of waste materials. However, a modern perception of CE at different levels (micro, meso, and macro) indicates that CE is rather a systemic tool to achieve sustainability and a new eco-effective approach of returning and maintaining waste in the production processes by closing the loop of materials. In this frame, CE and sustainability can be seen as a multidimensional concept based on a variety of scientific disciplines (e.g., engineering, economics, environmental sciences, social sciences). Nevertheless, the interconnections and synergies among the scientific disciplines have been rarely investigated in depth.

One significant goal of the book series is to study and highlight the growing theoretical links of CE and sustainability at different scales and levels, to investigate the synergies between the two concepts and to analyze and present its realization through strategies, policies, business models, entrepreneurship, financial instruments and technologies. Thus, the book series provides a new platform for CE and sustainability research and case studies and relevant scientific discussion towards new system-wide solutions.

Specific topics that fall within the scope of the series include, but are not limited to, studies that investigate the systemic, integrated approach of CE and sustainability across different levels and its expression and realization in different disciplines and fields such as business models, economics, consumer services and behaviour, the Internet of Things, product design, sustainable consumption & production, bio-economy, environmental accounting, industrial ecology, industrial symbiosis, resource recovery, ecosystem services, circular water economy, circular cities, nature-based solutions, waste management, renewable energy, circular materials, life cycle assessment, strong sustainability, and environmental education, among others.

Jayakrishna Kandasamy • Vimal K E K
Manavalan Ethirajan • Nishal Murali

Circular Supply Chain Management

A Sustainability Approach

Springer

Jayakrishna Kandasamy
School of Mechanical Engineering
Vellore Institute of Technology
Vellore, Tamil Nadu, India

Manavalan Ethirajan
Senior Delivery Manager
Kinaxis India Private Ltd.
Chennai, Tamil Nadu, India

Vimal K E K
Department of Production Engineering
National Institute of Technology
Tiruchirappalli
Tiruchirappalli, Tamil Nadu, India

Nishal Murali
Department of Mechanical Engineering
Sri Venkateswara College of Engineering
Chennai, Tamil Nadu, India

ISSN 2731-5509 ISSN 2731-5517 (electronic)
Circular Economy and Sustainability
ISBN 978-3-031-81749-6 ISBN 978-3-031-81750-2 (eBook)
https://doi.org/10.1007/978-3-031-81750-2

This Springer imprint is published by the registered company Springer Nature Switzerland AG
The registered company address is: Gewerbestrasse 11, 6330 Cham, Switzerland

We dedicate this book to our family and friends,
for their love and support
Thank You.

Preface

This book provides the fundamental knowledge on the circular economy and circular supply chain management. It also discusses the theories that have been developed or adopted in CE research. As traditional supply chains organizations have felt the importance of transforming their activities more circular, this book provides the need for adopting circular supply chain practices and the circular business models they can adopt. The barriers that the supply chains organizations will come across in the process of transformation and the drivers that they can focus to overcome the barriers are well elaborated. The key factors that help in decision-making process and modalities involved in modeling the circular chains are intricated well for the benefit of managers and other stakeholders. The role of I4.0 technologies in future supply circular chains is elucidated. This book also offers solutions to achieve the sustainable development goals by the supply chain organizations. Use cases and illustrations are provided for easy understanding of theory to practice.

Chapter 1 addresses the importance of incorporating circular economy (CE) principles into sustainable development frameworks and translating theoretical goals into practical implementation strategies. It presents widely adopted CE theories, supported by real-world examples, to guide future research. **Chapter 2** defines circular supply chain management (CSCM) and explores its significance and necessity.

The circular economy is a theoretical framework that seeks to establish an industrial system designed to regenerate resources and create a competitive advantage. **Chapter 3** explores specific practices that can be adopted to transform traditional supply chains into circular ones. **Chapter 4** identifies the obstacles organizations may face during this transition and provides strategies for overcoming them. To gauge their progress in moving from linear to circular supply chains, organizations require effective business models. **Chapter 5** offers an overview of contemporary circular business frameworks that can serve as benchmarks for this transformation.

Chapter 6 provides a detailed analysis of implementing circular operations, using decision models and analytical tools. **Chapter 7** addresses the question of how mathematical models can be used to optimize supply chain networks, transportation within circular networks, and decision-making processes that balance resilience and circularity. It also includes illustrative examples.

Chapter 8 explores different decision-making methods for assessing risks and selecting suppliers or partners within circular supply chain management. It also discusses the potential role of Industry 4.0 technologies in shaping future circular supply chains. **Chapter 9** addresses how these disruptive technologies can enhance and facilitate circular supply chain management. **Chapter 10** highlights the connection between sustainable development goals and CSCM, demonstrating how adopting circular practices contributes to achieving these goals.

Chapter 11 presents a real-world case study of a symbiotic relationship among paper, cement, and spirit production industries. It explores how these industries can work together to achieve circularity and highlights the challenges they may face. **Chapter 12** discusses another real-world example, focusing on the manufacturing and recycling of electrical components.

Overall, the book offers valuable insights for researchers, practitioners, and policymakers interested in advancing sustainable and circular supply chains.

Vellore, Tamil Nadu, India — Jayakrishna Kandasamy

Tiruchirappalli, Tamil Nadu, India — Vimal K E K

Chennai, Tamil Nadu, India — Manavalan Ethirajan

Chennai, Tamil Nadu, India — Nishal Murali

Acknowledgments

The authors would like to acknowledge the help of all who supported as in this project. Without their support, this book would not have become a reality.

Jayakrishna Kandasamy
Vimal K E K
M. Nishal
Manavalan Ethirajan

Contents

About the Authors

Jayakrishna Kandasamy is a professor in the School of Mechanical Engineering at the Vellore Institute of Technology University, India. Jayakrishna's research is focused on the design and management of manufacturing systems and supply chains to enhance efficiency, productivity, and sustainability performance. More recent research is in the area of developing tools and techniques to enable value creation through sustainable manufacturing, including methods to facilitate more sustainable product design for closed-loop material flow in industrial symbiotic setup, and developing sustainable products using hybrid biocomposites. He has mentored doctoral students and graduate and undergraduate students which have so far led to 104 journal publications in leading SCI/ SCOPUS Indexed journals, 46 book chapters, 92 refereed conference proceedings, 4 authored books, and 12 edited books in CRC or Springer Series. Dr. Jayakrishna's team has received numerous awards in recognition for the quality of the work that has been produced. He teaches undergraduate and graduate courses in the manufacturing and industrial systems area and his initiatives to improve teaching effectiveness have been recognized through national awards. Notably, he is ranked among the Top 2% most influential researchers globally in the prestigious 2023 and 2024 Stanford University World Ranking. He has also been awarded Global Engineering Education Award from Industrial Engineering and Operations Management (IEOM) Society International, USA, in 2021; Institution of Engineers (India)-Young Engineer Award in 2019; Distinguished Researcher Award in the field of Sustainable Systems Engineering in 2019 by International Institute of Organized Research; and Best Faculty Researcher Award for the year(s) 2016–2021 consecutively. He is a coordinator of Circular Economy Club, VIT University; Academic Editor of *Mathematical Problems in Engineering*, Wiley-Hindawi Publications; and Editorial Board Member of *Journal of Operational Research for Engineering Management Studies* (JOREMS) and *Green and Low-Carbon Economy*; Area Editor of *Operations Management Research*; Associate Editor of *Circular Economy*, *Frontiers in Sustainability*; and Book Series Editor of Industrial Engineering, Systems, and Management, CRC Press.

Vimal K E K received his Ph.D. degree in Sustainable Manufacturing in 2016 and M.Tech degree in Industrial Engineering in 2012 from National Institute of Technology Tiruchirappalli, India, and Production Engineering degree in 2010, from the PSG College of Technology, India. He has more than 12 years of teaching and research experience in Mechanical Engineering with special emphasis in supply chain management. He has worked as Guest Editor of the *International Journal of Enterprise Network Management* and Polymer and Polymer Composites. Presently, acts as a reviewer for more than 30 prestigious Web of Science journals. In addition, he has also published as an author (and co-author) 4 books with international reputed publishers, 20 book chapters, and more than 100 articles in journals and conferences. He is one of the top researchers working in the area of Circular Economy and Life Cycle Assessment in India. He has delivered over 30 guest lectures in various Faculty Development Programmes (FDPs) (within India and abroad) and conferences organized by reputed institutions. He has acted as session chair in many reputed international conferences. He has organized two Scopus indexed international conferences. He is currently working as assistant professor, Department of Production Engineering, National Institute of Technology, Tiruchirappalli, India.

Manavalan Ethirajan is a Senior Delivery Manager at Kinaxis India Pvt. Ltd, specializing in product development, project management, and supply chain solutions with over 18 years of experience. He manages high-budget international projects, delivering Enterprise Resource Planning (ERP), and supply chain management solutions, and has expertise in tools like Kinaxis Rapid Response, Oracle ERP, and James Donald Armstrong (JDA) Software. He holds a Ph.D. in Digital Supply Chain from VIT, along with multiple certifications, and has worked with companies like Oracle, Blue Yonder, and Gilead Sciences. Dr. Ethirajan is skilled in project delivery, solution design, and cross-functional team management across diverse industries.

Nishal Murali is an assistant professor at Sri Venkateswara College of Engineering, with over eight years of experience in both industry and academia. He holds a master's degree in Manufacturing Systems Engineering from Coventry University, UK, and a bachelor's degree in Mechanical Engineering from Panimalar Engineering College, Anna University. His research focuses on sustainable management, lean manufacturing, sustainable supply chain management, and quality management. His contributions extend to areas such as supply chain management, maintenance engineering, and ergonomics. He has authored and co-authored nine book chapters and published seven research papers. He has published three patents and two patents have been granted.

Abbreviations

Chapter 1: Circular Supply Chain Management: An Introduction

CE	Circular Economy
CSCM	Circular Supply Chain Management
SC	Supply Chain
CLSC	Closed Loop Supply Chains
SCM	Supply Chain Management
OLSC	Open-Loop Supply Chains

Chapter 2: Closing the Loop Through Circular Supply Chain

CSC	Circular Supply Chain
CE	Circular Economy
SCM	Supply Chain Management

Chapter 3: Transitioning of Traditional Supply Chain to Circular Supply Chain

KPI	Key Performance Indicators
CSM	Circular Supply Chain

Chapter 4: Barriers of Circular Supply Chain

CSC Circular Supply Chains
CE Circular Economy
SME Small and Medium-Sized Enterprises
KPI Key Performance Indicators

Chapter 5: The Smart Manufacturing Concept

APSRG All-Party Parliamentary Sustainable Resource Group
CBM Circular Business Models
CSCM Circular Supply Chain Management
SSCM Sustainable Supply Chain Management
GSCM Green Supply Chain Management
CCM Circular Supply Chain Management
SBM Sustainable Business Models
MRO Maintenance, Repair, Overhaul
COI Collaborative Innovation
NPD New Product Developments
CE Circular Economy
BMICE Business Model Innovation for Circular Economy
PSS Product-Service Solutions

Chapter 6: Understanding the Risk Associated in Implementing Circular Economy

CE Circular Economy
CSR Corporate Social Responsibility
PCB Printed Circuit Board
GHG Greenhouse Gas
GWh Gigawatt Hour
CEO Chief Executive Officer
IT Income Tax
SWOT Strengths, Weaknesses, Opportunities, and Threats
R&D Research and Development
IP Intellectual Property
JV Joint Venture
CSR Corporate Social Responsibility

Chapter 7: Circular Supply Chain Business Models to Incorporate Circularity Concept

PLE	Extending Product Life
DFR	Designing for Recycling
RPO	Remanufacturing Product Ownership

Chapter 8: Decision Making Models in Circular Supply Chain

CSCM	Circular Supply Chain Management
LCA	Life Cycle Assessments
SCM	Supply Chain Management
MO-MILP	Multi-Objective Mixed Integer Linear Programming Model
LCA	Life Cycle Assessment
EPDs	Environmental Product Declarations

Chapter 9: Role of Industry 4.0 in the Future Circular Supply Chain Management

CPS	Cyber-Physical System
RFID	Radio-Frequency Identification
DCE	Digital Circular Economy
UAVs	Unmanned Aerial Vehicle
IFR	International Federation of Robots
IIoT	Industrial Internet of Things

Chapter 10: Closing the Loop: Unlocking Sustainable Development Goals Through Circular Supply Chains

CSCM	Circular Supply Chain Management
CE	Circular Economy
SDG	Sustainable Development Goals
CSC	Circular Supply Chains
CSCM	Circular Supply Chain Management
NPI	New Product Introduction
SCV	Supply Chain Visibility
PLM	Product Lifecycle Management

Chapter 11: Investigating Circular Supply Chain Strategies and Industry 4.0 Implementation: A Case Analysis of a Paper Manufacturing Company

MGFA	Multi-Grade Fuzzy Approach
CSCM	Circular Supply Chain Management
CSR	Corporate Social Responsibility
SME	Subject Matter Experts
SPSS	Statistical Package for the Social Sciences
KPI	Key Performance Indicators
ICT	Information and Communication Technology

Chapter 12: Exploring Circular Supply Chain Practices and Industry 4.0 Integration: A Case Study in a Pump Manufacturing Organization

CE	Circular Economy
MGFA	Multi-Grade Fuzzy Approach
CSCM	Circular Supply Chain Management
SCM	Supply Chain Management
SME	Subject Matter Experts
ICI	Circularity Index
PF	Promotional Factors

Chapter 1
Circular Supply Chain Management: An Introduction

Abstract Global value chains have emerged and international trade has increased dramatically as a result of globalization. The notion of circular economy (CE) has attracted substantial interest recently among scholars, practitioners, and regulators due to the tremendous waste generated at all phases of a product life cycle. CE acts as alternate answers to the current linear economic model of manufacturing and consumption. Circular supply chain management (CSCM) is the term used to describe the integration of the CE into Supply Chain management. In light of this, solutions utilizing industry 4.0 technologies are recognized as essential to the effective advancement of the circular economy. Industries are focusing more towards circular supply chain as it has many benefits for all with a downside to financial investment. Many organisations and research institutes are coming up with a framework to implement Industry 4.0 in their supply chain. In this chapter the importance and impact of CSCM and their importance in achieving sustainability is discussed.

Keywords Circular economy · Circular supply chain management · Industry 4.0 · Globalisation · Linear supply chain and sustainability

1.1 Introduction

Research on the relationships between the expansion of the economic system and the problems of environmental degradation has increased over the twentieth century. Voices opposing globalization and emphasizing the necessity of a green transition are becoming more prevalent (Ghorbani et al. 2023). While it would be excessively optimistic to expect environmental sustainability to take precedence over other considerations in the context of an economic rebound, it is unavoidable that demand- and supply-side players across a wide range of industries embark on a more environmentally friendly course as soon as feasible and pursue resilient, sustainable supply chains. The twenty-first century's technology advancements and theoretical understanding must be applied in tandem to achieve the economic, social, and environmental elements. This calls on businesses to deal with the

J. Kandasamy et al., *Circular Supply Chain Management*, Circular Economy and Sustainability, https://doi.org/10.1007/978-3-031-81750-2_1

limitations of natural resources, the need to manage them prudently, and the deliberate participation of both families and institutions. It is more crucial than ever to stress the necessity of altering mindsets if sustainability initiatives are to be successful. Not only must consumer society be rethought, but techniques that can preserve existing production in full or in part must also be developed, all while requiring less material and energy.

Figure 1.1 represents the actors involved in a circular economy. When considering the entirety of contemporary economic processes and examining trends in consumption patterns, it is evident that resource usage is steadily rising. About half of the consequences of climate change and over 90% of biodiversity loss and water pollution are caused by the worldwide mining and processing of natural resources (Wassie 2020). An increasing number of complicated goods are being made, and an increasing number of materials are being employed both quantitatively and qualitatively, due to the diversity of materials' continuous and rapid increase. The annual global raw material extraction tripled between 1970 and 2017 and has continued to grow since then, according to the European Commission's. This information is based on data from Circle Economy Strategic Alliances from 2015. The paper "Our world without waste?" summarizes global material flows and the intricate processes that underlie them as follows. Every year, we turn 102.3 billion metric tons of raw materials into products. 93 billion tons of raw materials (including minerals, ores,

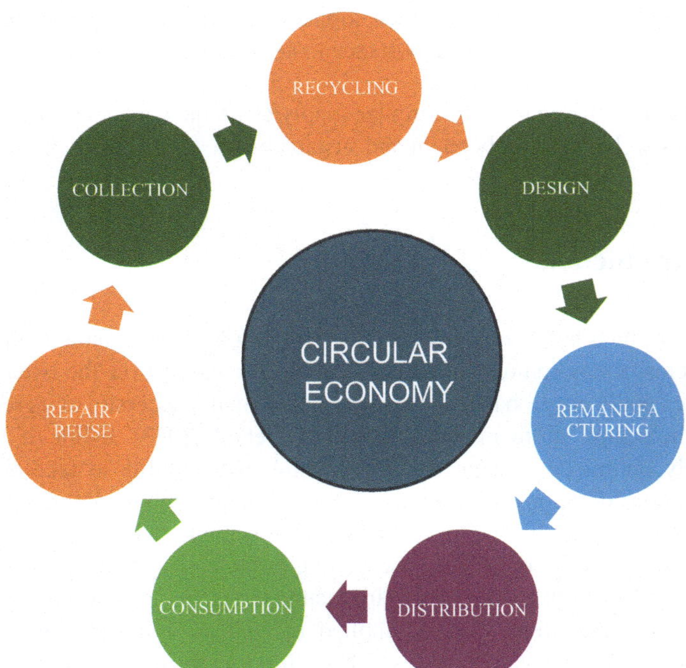

Fig. 1.1 Circular economy

fossil fuels, and biomass) were extracted from the natural environment (Agrawal et al. 2023). Of these, 25% are utilized to create durable commodities such as automobiles, homes, and other items, while the remaining 9.3 billion tons or less are returned to the economy. 2.67 billion tons, or two-thirds of the materials moving through the economy, were emitted as trash or, for example, as carbon dioxide from fossil fuels at the conclusion of the process, while 2.3.7 billion tons were "durable materials," such as end-of-process structures and infrastructure. These materials are reused through composting, recycling, soil enhancement, biogas generation, and wastewater treatment. As a result, every year, mankind produces goods from about 100 billion tons of raw resources (Nandy et al. 2022). In this aspect we need to improve asset efficiency in the fundamental necessities of food supply, living, housing, and transportation/mobility are the key drivers of society's resource and carbon footprint (Bhattarai et al. 2021).

Four categories of raw materials—minerals, ores, fossil fuels, and materials of biological origin—are involved in the material flow operations of the global economy, according to the UN technical document on the subject. This is expected to escalate in the ensuing decades; by 2050 (Dold 2020), resource extraction using the present production and consumption patterns may have doubled. It is anticipated that materials and minerals with biological origins will develop at the greatest rate. The way resources are used globally is not just growing but also varying by area. All other areas were net exporters in 2000, with the exception of high-income nations. Currently, one billion tons of material are exported by high-income nations, with Australia and the United States leading the way in this growth (Schandl et al. 2020). Approximately 750 million tons are exported by upper-middle-income nations. The Pacific region's material footprint was expected to be 11.4 tons per person in 2017. The ultimate material usage per person in North America is thirty tons. In Europe, it is 20.6 tons, but in every other continent, it is less than 10 tons. However, if they are extensively implemented in international supply chains, supply chains that prioritize sustainability issues and are especially built on the circular economic model might contribute to the more positive development of these trends. Given the aforementioned procedures and considerations of supply chain sustainability, intervention is obviously necessary. A number of nations, areas, and industrial firms are focusing on creating long-term plans. Many techniques have arisen in this area, following worldwide trends: material cycle definitions, cross-industry networking, industrial symbiosis approaches, waste hierarchy-based definitions, resource-oriented definitions, and economy-focused definitions. It is obvious that there is a greater need than ever to reconsider global supply chains and consumer trends. The circular economy model might be one method for this.

The circular economy, which is an umbrella term for a variant of economic, social, and environmental processes and bond of influence with sustainability in mind, is a sort of innovation framework formed by combining earlier ideas (Merino-Saum et al. 2020). It provides an overview of the theories that have evolved in the last several decades in the fields of ecology and environmental economics. Global and local supply chains must take immediate, thorough, and coordinated action to address sustainability concerns. This is why the European Commission's new circular economy action plan, which will be unveiled in March 2020, is a step in the right direction. The new strategy intends to expedite the transition mandated by the European Green Deal and expand the circular economy from trailblazers to significant economic actors. In light of this, the Commission designated the following sectors as critical value chains: electronics, ICT (information and communication technologies); automobiles and batteries; packaging; plastics; textile industries; building and construction; food; water; and nutrition. Supply networks are undoubtedly moved toward sustainability and the circular economy model by concentrating on these aspects. Since the late 1990s, supply chain management (SCM) has been recognized as a multidisciplinary area of study that has been developing and continuously looking for solutions to issues that arise now. A product, service, or piece of information that is valuable to customers and other stakeholders is created through the integration of critical processes from the producer to the end user through the activity of supply chain management.

The supply chain process encompasses a number of product-related services in addition to the derivation of natural materials and the transfer of completed goods to end users. Manufacturing, marketing, and finance, as well as strategic resource planning, business process interactions, risk sharing, and supply providers' engagement in product development, are all part of the supply chain, which is based on information systems (Islami 2023). Supply chain management improves living standards, lowers pollution and energy use, and generates jobs in a healthy community. The people population, economic growth, and living standards are all steadily rising, which is directly related to overuse of resources and environmental deterioration. In this regard, one of the biggest problems confronting mankind is developing more sustainable supply chains. Events in the last few decades have occasionally brought attention to how vulnerable global supply systems are.

The supply chain was seen as a star-shaped structure driven by vertical, dyadic customer-supplier connections throughout the 1970s, when the growth pole hypothesis predominated. The mutual trust between first-tier suppliers and customers contributed to the supply chain's transformation into a pyramid-shaped structure in the 1980s. Information and knowledge exchange techniques have been shown to have a favorable impact on mutually beneficial interactions between suppliers and buyers/customers. The supply chain literature of the 1990s focused mostly on customer-led initiatives and different competitive and collaborative alliances between businesses. Thanks to the numerous climate change agreement and the standards and regulations they establish, environmental consciousness has increased throughout the region in the 2000s. All of this is important for supply chains as well as for individual businesses. Over the past ten years, supply chain research directions have changed due to the increasing value of environmental and climate protection, as well as the demand to achieve existing goals and requirements. These have been discovered via environmental management techniques, sustainability concerns, and the real-world implementation of the fresh opportunities provided by the circular economy and green supply chains.

It is important to concentrate on the manufacturing supply chain's performance from the producer to the end user (Jha et al. 2022). The crucial implementation of supply chain management with circularity in industrial processes aligns with this demand. In contrast to closed-loop supply chains (CLSCs), circular economy businesses can work together with businesses in related and unrelated industries to maximize salvage value and achieve zero waste. However, given that the area of circular economy is relatively new, it was found that there had not been enough discussion of the CSCM problem in the literature (Foroozanfar et al. 2022). Malaysia's industrial industry has not given CSCM much thought, as seen by the lack of proactive measures to recycle or recover returns in circular contexts. Environmentalists are worried that a significant amount of the stuff that is gathered is burnt rather than recycled. The integration of sustainability concepts into supply chain management (SCM) has been expressed through the introduction and interchangeability of a number of concepts, including open-loop supply chains (OLSC), closed-loop supply chains (CLSC), green supply chains, and environmental supply chains.

Although these ideas show different levels of integration of sustainable supply chains, the majority of them lack the circular viewpoints that are considered essential to the basis of CSCM models. The study began by examining the definition via a methodical review of the literature in order to gain a comprehensive understanding of the current status of CSCM research. The results indicated that the field has a lot of promise and that more research should be done using the CSCM conceptualization, which includes supply chain functions (reorientation), appropriate business models (closed and open loop), and restorative and regenerative processes in order to achieve a zero-waste goal in the context of the circular economy. Moreover, the supply chain successfully facilitated the establishment of materials recovery ecosystems in the packaging industry using the circular supply chain architecture. The study found that such ecosystems are more complex than normal linear value chain systems because they need the inclusion of additional players and processes that execute circular flows of materials.

Figure 1.2 shows the interlink between sustainability and the level of changes that it makes. As of right now, the CSCM is only an idea, although a potentially effective one. Large-scale implementation of the shift by manufacturing companies to greater CSCM is still pending (Amjad et al. 2021). If this acceptance is sluggish, it will be disastrous, particularly for emerging nations like Malaysia that have an abundance of waste resources that may be converted into useful materials for production. Moreover, an analysis of the body of existing literature reveals that there is currently a dearth of comprehensive perspectives on CSCM. This makes it difficult to differentiate it from other supply chain sustainability ideas and impedes further development in the area. Therefore, by clearly defining the CSCM idea and activities, the study seeks to construct a conceptual model.

Global production, consumption, and trade patterns are still, however, gravely unsustainable. If there is no change in the way goods are obtained, produced, distributed, consumed, recovered, and regenerated, the world will eventually run out of many natural resources at its current level of consumption. CE, which is gaining popularity as a superior theory to the conventional linear (take, manufacture, and discard) economic model, is one significant ideology that might lead to this shift (Grafström and Aasma 2021). In the literature and in practice, the CE concept is

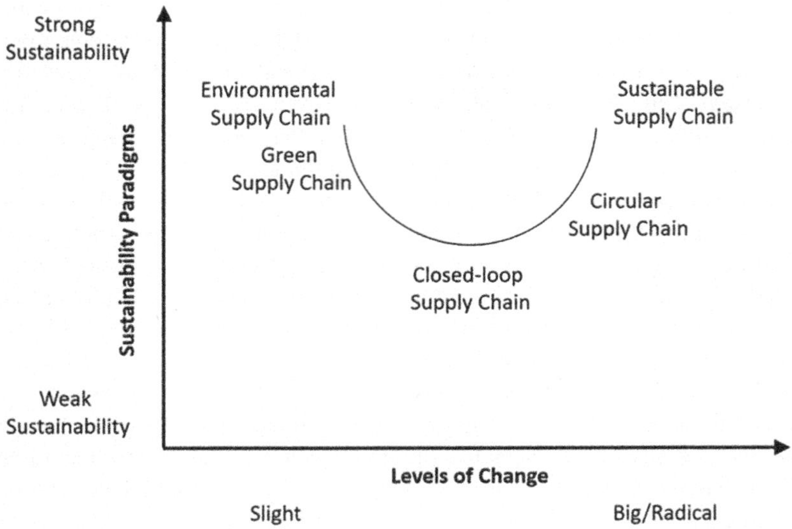

Fig. 1.2 Features of the SCM

becoming a powerful force behind sustainability. It is also starting to be acknowledged as having a significant potential to assist businesses in making significant progress toward sustainable performance.

1.2 Background Information in History

Two long-standing threads make up the concept of the circular economy: the one addresses the flow of materials within an economy, while the second takes into account the possible economic conditions that may encourage this flow. Since the 1960s and 1970s, when the environmental movement was first emerging, these two schools of thought have coexisted peacefully. The materials strand originated from the idea of "industrial ecology," a term coined in economic geography in the 1940s to describe the factors that influence where industries are located in order to maximize resource utilization and minimize waste, along with the related term "industrial symbiosis". The industry has to shut the loop that goes from the user to the production in the next industrial revolution. The article's initial design would have included characteristics that would have made it easier for them to return and remake if American business were to take on the responsibility of ending this loop.

The media, business executives, and public officials all need to acknowledge and respect them. They need to be ingrained in society and accepted by both the government and business. Obviously, there hasn't been a wider spread of industrial ecology ideas yet, but at least academics have interpreted to activeness seriously. The arrangement and synchronization of marketing, sales, R&D, production, logistics,

IT, finance, and customer service organizational operations both inside and between business units and organizations in order to enhance system efficiency and productivity and provide competitive benefits. An appealing viewpoint is provided by circular supply chain management (CSCM), which incorporates the idea of a zero-waste economy as well as restorative and regenerative cycles created using circular thinking (Maguire and Robson 2023). A wide variety of stakeholders must be included in the adaptation of the business model and supply chain designs in order to realize the goal of zero waste through the restoration and regeneration of materials.

1.3 The Necessity of CSM

Reusing seemingly wasted materials is made possible by a circular supply chain, which also seeks to turn product waste into new goods. Compared to more conventional linear supply chains, where producers dispose of product waste, this is a modification. Opportunities to improve sustainability and the environment are provided by circular supply networks. Manufacturers have been compelled to find creative ways to sustain manufacturing since 2020 because to growing prices and limited access to raw resources. This has led to a shift from outdated methods to new ones that provide cost savings and cleaner supply chains. The looping process helps the manufacturing process by reducing production and transportation expenses as well as the quantity of material that is thrown out into the environment. Less waste, more income, and a less environmental effect are guaranteed by the outcomes. Everyone wins in this scenario. The closed-looped system, depicted in this model, gives businesses that must adhere to stringent trash and recycling laws an edge over rivals and access to customers who choose eco-friendly goods (Singh and Joshi 2024).

The circular economy idea also offers chances to use existing infrastructure and equipment in new ways and for longer periods of time, which minimizes emissions and maximizes outputs. Although switching from the linear models is necessary for this holistic approach, it is well worth the effort because it lessens regulatory requirements and helps to avoid potential fines from regulatory oversight. A novel approach to procedures that eventually result in new goods, reduced costs, new income streams, and customer growth while supporting green principles may be ensured by effective planning, evaluating, analyzing, and supervising. New ideas are needed in the modern day, and recycling helps the economy and enterprises by adding several advantages to the supply chain and generating more efficient techniques when resources are few.

Using resources as long as feasible and recovering goods at the end of their life cycle to recycle them into new ones are key components of a circular supply chain. The circular supply chain management approach is a more environmentally friendly option than the traditional linear model because of its restorative qualities. The circular economy, sometimes referred to as reverse supply chain management, has gained a lot of interest recently as a result of the safety guidelines established by local governments to reduce their carbon footprint. Additionally, businesses have realized that connecting the conclusion of a supply chain to its beginning results in increased profit and less waste.

1.3.1 Advantages of a Circular Supply Chain for a Business

1.3.1.1 Minimal Production Expense

Remanufacturing waste items allows a corporation to minimize production expenses since circular supply chain management reuses the materials. Since the majority of your need will be satisfied by the reusable items you recover, you do not need to constantly purchase new raw materials.

1.3.1.2 An Environmentally Friendly Strategy

The goal of the circular supply chain is to minimize waste, if not completely eliminate it, in order to keep it out of landfills. With this approach, the businesses may adopt sustainability since the "waste" is transformed into an environmentally friendly resource. By reducing trash and reusing as much as you can, you are contributing to environmental preservation by employing this strategy.

1.3.1.3 Following Official Directives

Environmental challenges are prompting governments worldwide to enact increasingly stringent regulations. As a result, businesses are compelled to adopt environmentally friendly practices and prioritize recycling to comply with the laws and regulations being implemented. Violating these restrictions can lead to significant fines, which not only impact a company's earnings but also tarnish its reputation. Implementing a circular supply chain can help businesses adhere to these rules and avoid being subjected to restrictions, ensuring compliance while maintaining operational efficiency.

1.3.1.4 Defence Against Changes in Price

In today's unpredictable world, various factors can disrupt business operations or make it challenging to procure raw materials, such as increasing prices. However, employing a circular supply chain strategy can shield businesses from such instability. By understanding the amount of raw material generated through product recycling, companies can accurately estimate the resources required for creating new items. This enables them to mitigate the impact of fluctuating prices and ensure a more stable supply chain.

1.3.1.5 Enhanced Client Allegiance

With an increasing environmental consciousness among consumers, there's a growing emphasis on considering the environmental impact of products before making purchases. Many consumers are willing to pay a premium for sustainable items, recognizing their benefits for the environment despite their higher cost compared to traditional products. Embracing the circular supply chain approach can bolster a company's reputation with customers, demonstrating a commitment to sustainability and responsible environmental practices. This, in turn, can foster greater loyalty among customers, encouraging continued business relationships.

1.4 Approaching CSM

Environmentalists and academics are calling for the quick adoption of sustainable methods that will cut waste and carbon emissions. This cry is becoming louder. But a lot of supply chain executives and companies have been sluggish to respond. Only when companies adopt a comprehensive approach will the circular supply chain prove effective. This is how supply chain-wide circularity may be attained.

1.4.1 Procurement

Corporate strategy may incorporate sustainable principles through the use of circular buying. Procurement teams may speed up the implementation of circular business models by shifting demand towards sustainable items. Supplier selection, for instance, may provide preference to vendors that demonstrate a strong commitment to sustainable practices. Furthermore, purchasers can set rules emphasizing dealing with manufacturers who employ recycled, recyclable, or sustainably sourced goods. While new circular supplier frameworks exist, the engagement between buyers and providers is crucial. Sustaining existing connections through adaptation will be a major motivator for altering habits. Product Design The development of circular supply chains will be greatly aided by advances in product design. Designers can seal the loop at both ends. Firstly, by using bio-based or recyclable components in products, so lessening the burden on our natural resources. Secondly, circular design may rethink the product life cycle by developing modular objects that are easy to repair and disassemble and, thus, appropriate for re- and upcycling, as can be observed in some of the real-world circular economy instances.

1.4.2 Manufacturing

With sustainability objectives a rising focus for industrial organizations, the circular supply chain is on the table. However, Bringing the idea of the circular economy to existence within the business enterprise value chain involves significant relation in core group industry and logistics. Minimizing waste, carbon emissions, and energy usage through streamlined and efficient production procedures is vital. Embracing this innovation will demand investment. However, it will deliver large returns through materials cost reductions and result in better sales as customers demand greener products. Furthermore, industrial business models must evolve to adopt closed-loop systems spanning waste and resource management. By lowering dependency on new resources, producers may reuse and recycle as much as feasible. Finally, education is a crucial pillar in the transition from a linear to a circular paradigm. Businesses must spend in training staff about circular economy ideas and best practices to guarantee a seamless transition.

1.4.3 Logistics and Distribution

Logistics and distribution are among the major sources to carbon emissions. While most supply chains strive toward efficiency as a driver of profit, the circular economy will demand extra objectives and targets connected to sustainability standards. Some of the improvements for logistics and distribution teams will entail embracing green mobility, such electric vehicles (Qadir et al. 2024). Additionally, more complex route planning should be employed to cut out inefficiencies and minimize emissions. Reverse logistics are another major topic. Establishing networks and mechanisms to return items from consumers to their point of origin will assist repair, recycling, refurbishing, replacement, and redistribution. Additionally, it can aid with clean disposal and dismantling of products so raw materials can be reused.

Other efforts to minimize transport emissions should entail collaboration with partners to build localized manufacturing and distribution networks. Finally, supply chains may be shortened by business strategies that encourage local production. The ultimate result of these measures will be to lessen the effect and emissions associated with long-distance transportation while simultaneously bolstering local economies.

1.4.4 Marketing and Sales

Teams in charge of sales and marketing may accelerate the shift to a circular model by influencing behavior, culture, and attitudes. From a commercial perspective, the procedure will entail informing clients about sustainable goods and services in an efficient manner. Promoting advantages, financial savings, beneficial effects on the environment, and other points might encourage customers to support sustainability.

1.5 CSM Theories

1.5.1 The Ecological Economy

"*In a circular economy, the value of products and materials is preserved for as long as possible, waste and resource consumption are minimized, and resources are retained within the economy when a product reaches the end of its useful life, to be reused indefinitely to create additional value*" is how the European Commission defines the shift to a circular economy. While the concept of the circular economy is widely accepted, there's a pressing need for practical application due to the sub-par performance of current circular economy initiatives. Leveraging manufacturing processes through component optimization using technology and modeling can enhance product quality and reduce waste generation (Qadir et al. 2024). Additionally, integrating various participants within the supply chain can bolster production capacities in reverse supply chains, including remanufacturing, product recovery, and reverse logistics. Such integration holds the potential to expedite the transition of manufacturing toward circular economy practices.

1.5.2 Management of a Circular Supply Chain

Circular supply chain management involves applying circular perspectives to the management of the supply chain and the industrial and natural environments in which it operates. The goal is to achieve a zero-waste target through system-wide innovation in business models and supply chain operations that encompass the entire product/service life cycle, from design to end-of-life and waste management, involving all stakeholders. Interestingly, CSCM recognizes two types of product circularity: open-loop supply chain (OLSC) and closed-loop supply chain (CLSC), which mirror the CSCM components.

1.5.3 Supply Chains in Closed Loops

Supply Chains in Closed Loops (CLSC) is the process of re-establishing the connection between the initial forward supply chain and the reverse supply chain in order to bring returns back into downstream production and distribution systems

(Theyel and Hofmann 2021). The loop can link to some location on the onward supply chain; it is not required to be connected to the manufacturing location. It refers to the ultimate location of goods at the conclusion of the reverse supply chain, where companies may decide whether to close the loop or keep it open. Reclaiming end-of-use products instead of discarding them is a component of the CLSC concept, which have been shown to improve supply chain sustainability. The enterprises old supply chain management was replaced with a green supply chain management model that relies on closed-loop logistics and recycling to transfer materials throughout supply networks.

1.5.4 Supply Chain Open-Loop

On the other hand, Supply Chain Open-loop (OLSC) may also be used as the destination of the returns in the reverse flow. When it comes to managing common reverse logistics activities, the capacity to connect return operations with original operations is often what separates CLSCs from OLSCs. Products enter and leave the supply chain in an OLSC in a single-direction configuration at two independent sites. The products in OLSCs are repossessed by other parties who are able and willing to reuse them instead of having to send them back to the original makers. For instance, used beverage cans (UBC) were obtained by Novelis, a recycling business in the UK, and then turned into can body coils that were sold to other can producers. The majority of Tetra Pak's circular supply chain in China is made up of open loops. Recycling companies are essential to Tetra Pak's circular supply chain because the business does not recycle UBCs. This is not the case with CLSCs, where items are sent back to the original manufacturers for additional salvaging operations in order to utilize the "available" value through reuse and remanufacturing procedures. Moreover, there are two kinds of resource flows seen in both closed and open circular supply networks: forward-flow chains and reverse-flow chains. Items typically flow forward from producers to consumers, but items that flow backward do so through with recovery procedures such as recycling, reusing, remanufacturing, and refurbishing (Gunasekara et al. 2023).

1.5.5 Supply Chains That Go Forward

Products are moved from the manufacturer to the customers generally through the forward supply chain, it is comprised of system of integrated facilities and supply alternative that buy raw materials, process them into semi- and completed products, and then deliver the created goods to consumers. More precisely, the forward supply chain is made up of several processes that convert natural materials into final goods. This states the managers overseeing an onward supply chain must improve the effectiveness of a number of critical operations, including production, purchasing, order fulfilment, and management. Several methods of improvement, such as corporate customer relationship management and the supplier development program, can be used to achieve this. In addition, a number of manufacturing techniques are implemented in the process to improve environmental performance, decrease waste creation, and boost eco-efficiency. Cleaner production, green design, remanufacturing, and lean manufacturing are a few of the successful tactics. Putting these ideas into practice may lower the cost of raw materials and improve energy efficiency throughout the manufacturing stage, both of which might eventually improve the firm's performance and competitiveness.

1.5.6 Supply Chains in Reverse

Reverse logistics, which handle product recovery, returns, and excess inventory in the market, has been extensively acknowledged in the literature as the primary element of circularity (Khan et al. 2023). Its contribution to commodity circularity in onward supply chains and its ability to reduce waste created through various recovery operations, including recycling, remanufacturing, and reuse, reverse logistics has also emerged as a key component of the supply chain. This has given rise to the idea that supply chain sustainability, green marketing, and green logistics all depend on reverse logistics. Backward logistics supports the supply chain by putting returns into circulation and converting them into marketable products, therefore minimizing waste throughout the process. Thus, controlling the reverse flow chain to maximize the value of product returns is the ideal strategy for businesses looking to achieve sustainability.

Reverse logistics purview has now surpassed assist parts management to include post-service, repair, remanufacturing, and other related fields. However, organizations find it challenging to adequately plan, manage, and oversee the process since it has not been mapped under the broad category of reverse logistics. Moreover, the majority of companies implemented reverse logistics initiatives via a department or group unrelated to logistics. The current state of affairs has resulted in additional complexities that may pose challenges to the whole process of product circularity within supply chain management.

1.5.7 The Recovering Procedure

Operations in reverse supply chains are complex. Businesses must implement the recovery process in at least two distinct rework locations repair and refurbishment facilities and remanufacturing centers in order to get around this. Less money was invested in the repair and refurbishment centers, and typical refurbishing tasks are performed by people with human talents to restore returns to as good of condition as new items. Remanufacturing facilities, on the other hand, are focused on large financial investments and technology assistance to create better remanufactured items from product returns. These transformation systems' capabilities are available at the factory or at gathering centers, that are frequently located near to the customers.

1.5.8 Resale and Remarketing

Remarketing involves the sale, relocation, and storage of recovered goods in order to resell or reuse them (Kushwaha et al. 2022). In this case, a range of various marketing channels may be employed to sell recovered or used goods. Offering new products in the existing market while differentiating them from reconditioned ones is one way. A further means of spreading the word is by direct marketing to "specialty brokers". Following that, these brokers will distribute to both end users and other parties, such as retailers who provides reduced items. Because recovered, reused, and refurbished objects may have longer lifetimes than new ones, there is a reduced need for new resources, which is good for the environment. A number of factors, including as product life expectancy, degree of innovation, originality of design, customer demand, and market value of the products, were shown to influence the resale performance of reused things. Some managers are concerned about how the costs of reconditioned items will impact newly introduced products and designs on the market, especially in light of the disputes with OEM suppliers.

1.6 Conclusion

A key factor in manufacturing organizations' performance is supply chain management. Businesses may improve customer happiness, lower operating costs, boost production efficiency, and boost profitability by comprehending and managing the various supply chain components. Circular Supply Chain offers many advantages to industries like creating a sustainable product, improve the resiliency and creating less waste. Coupling the CE with a supply chain concept that an organization has put in place results in a circular supply chain, which offers a significant advantage to the sustainability of the supply chain operation. In this framework, CSCM calls for circularity in organizations and supply chain networks. More people need to be taught the ideas of system optimization and industrial ecology. In recent times, CSCM has become a more popular study area. It has a significant potential to raise manufacturing firms' three-dimensional sustainability efficiency. CSCM is an alternative to the take, make, consume, and dispose of manufacturing paradigm found in a linear supply chain. Supply chain operations are more effective when CSCM is implemented. Enhancing resource efficiency, supply chain effectiveness, growth in the economy, value propositions, end of life tactics, competitiveness, and other aspects of an organization's operations are just a few of the significant advantages that come with CSCM adoption.

References

Agrawal R, Bhagia S, Satlewal A, Ragauskas AJ (2023) Urban mining from biomass, brine, sewage sludge, phosphogypsum and e-waste for reducing the environmental pollution: current status of availability, potential, and technologies with a focus on LCA and TEA. Environ Res 224:115523. https://doi.org/10.1016/j.envres.2023.115523

Amjad MS, Rafique MZ, Khan MA (2021) Leveraging optimized and cleaner production through industry 4.0. Sustain Product Consum 26:859–871. https://doi.org/10.1016/j.spc.2021.01.001

Bhattarai K, Conway D, Bhattarai K, Conway D (2021) Impacts of economic growth, transportation, and tourism on the contemporary environment. In: Contemporary environmental problems in Nepal: geographic perspectives. Springer, Cham, pp 563–662. https://doi.org/10.1007/978-3-030-50168-6_7

Dold B (2020) Sourcing of critical elements and industrial minerals from mine waste–the final evolutionary step back to sustainability of humankind? J Geochem Explor 219:106638. https://doi.org/10.1016/j.gexplo.2020.106638

Foroozanfar MH, Imanipour N, Sajadi SM (2022) Integrating circular economy strategies and business models: a systematic literature review. J Entrep Emerg Econ 14(5):678–700. https://doi.org/10.1108/JEEE-10-2021-0411

Ghorbani Y, Zhang SE, Nwaila GT, Bourdeau JE, Rose DH (2023) Embracing a diverse approach to a globally inclusive green energy transition: moving beyond decarbonisation and recognising realistic carbon reduction strategies. J Clean Prod 434:140414. https://doi.org/10.1016/j.jclepro.2023.140414

Grafström J, Aasma S (2021) Breaking circular economy barriers. J Clean Prod 292:126002. https://doi.org/10.1016/j.jclepro.2021.126002

Gunasekara L, Robb DJ, Zhang A (2023) Used product acquisition, sorting and disposition for circular supply chains: literature review and research directions. Int J Prod Econ 260:108844. https://doi.org/10.1016/j.ijpe.2023.108844

Islami X (2023) Lean manufacturing and firms' financial performance: the role of strategic supplier partnership and information sharing. BIJ 30(9):2809–2831. https://doi.org/10.1108/BIJ-02-2022-0084

Jha A, Sharma RRK, Kumar V, Verma P (2022) Designing supply chain performance system: a strategic study on Indian manufacturing sector. Supply Chain Manag Int J 27(1):66–88. https://doi.org/10.1108/SCM-05-2020-0198

Khan SA, Laalaoui W, Hokal F, Tareq M, Ahmad L (2023) Connecting reverse logistics with circular economy in the context of industry 4.0. Kybernetes 52(12):6279–6320. https://doi.org/10.1002/bse.2479

Kushwaha S, Chan FT, Chakraborty K, Pratap S (2022) Collection and remanufacturing channels selection under a product take-back regulation with remanufacturing target. Int J Prod Res 60(24):7384–7410. https://doi.org/10.1080/00207543.2022.2072786

Maguire S, Robson I (2023) Supply chain management and the circular economy. In: Sustainable development through global circular economy practices. Emerald Publishing Limited, Leeds, pp 79–111. https://doi.org/10.1108/978-1-83753-590-320231004

Merino-Saum A, Clement J, Wyss R, Baldi MG (2020) Unpacking the green economy concept: a quantitative analysis of 140 definitions. J Clean Prod 242:118339. https://doi.org/10.1016/j.jclepro.2019.118339

Nandy S, Fortunato E, Martins R (2022) Green economy and waste management: an inevitable plan for materials science. Prog Nat Sci Mater Int 32(1):1–9. https://doi.org/10.1016/j.pnsc.2022.01.001

Qadir SA, Ahmad F, Al-Wahedi AMA, Iqbal A, Ali A (2024) Navigating the complex realities of electric vehicle adoption: a comprehensive study of government strategies, policies, and incentives. Energ Strat Rev 53:101379. https://doi.org/10.1016/j.esr.2024.101379

Schandl H, Lu Y, Che N, Newth D, West J, Frank S et al (2020) Shared socio-economic pathways and their implications for global materials use. Resour Conserv Recycl 160:104866. https://doi.org/10.1016/j.resconrec.2020.104866
Singh RK, Joshi HT (2024) Greening the supply chain: institutional pressures, data-driven innovations, and the road to circular sustainability. Bus Strateg Environ 33(6):6029–6044. https://doi.org/10.1002/bse.3788
Theyel G, Hofmann KH (2021) Manufacturing location decisions and organizational agility. Multinatl Bus Rev 29(2):166–188. https://doi.org/10.1108/MBR-02-2020-0035
Wassie SB (2020) Natural resource degradation tendencies in Ethiopia: a review. Environ Syst Res 9(1):1–29. https://doi.org/10.1186/s40068-020-00194-1

Chapter 2
Closing the Loop Through Circular Supply Chain

Abstract A circular supply chain (CSC) is a system that aims to eliminate waste and keep products and materials in use for as long as possible. Linear supply chain takes a linear path from raw material till the disposal. This increases the debris in the environment. A closed supply chain forms a loop from consumer to manufacturers which reduces the accumulation of debris. By design or intention, circular supply chains are among of the main facilitators of closing the loop for value recovery and maximizing profits. Although CSC is a new field, there isn't much discussion in academic circles about it from the perspective of a smoothly connected upward and reverse sc that aims to attain system-wide circularity. Circular economy increases the values of the materials when integrated into the supply chain. Green supply chain will help Manufacturers, consumers and nature to benefit together. Benefits of CSC are Low production cost, Environment-friendly approach, adherence to government guidelines, protection against price volatility, improved customer loyalty.

Keywords Circular supply chain · Remanufacturing · Circular economy · Refurbish · Recycle · Consumer

2.1 Introduction

The management of movement of raw materials, products and services is defined as Supply chain management. An effective way to reduce expense and lawsuits is by having a good supply chain management. In this modern world, industries use the linear model wherein the products are thrown away at the end of their life. This action does not go well with the environment as it disturbs the ecological balance. NGOs, industries and governments have shown concern about the global affect due to the disposal of different kinds of products. A circular supply chain is a system that aims to eliminate waste and keep products and materials in use for as long as possible (Orji and Ojadi 2023). This is in contrast to the traditional linear supply chain model, which is based on the "take-make-dispose" approach. This is also known as reverse supply chain management. Many companies like Kuebix, Amazon

J. Kandasamy et al., *Circular Supply Chain Management*, Circular Economy and Sustainability, https://doi.org/10.1007/978-3-031-81750-2_2

and Accenture follow the circular supply chain management which has greatly reduced the wastes are disposed into the environment. The CSCM decreases the carbon foot print by returning the used parts. The used parts are then repaired, recycled or reused. To have an eco-friendly future, industries must change their views and opinions on their products. At initial stage production methods, way of consumption and disposal of consumed products needs to change. These changes will not happen overnight but with time and patience it can be implemented. Reducing the waste has been major challenge for the industries and governments. Many industries started to move towards the green supply path as they are not able to come up with a solution for the increase in the cost of raw materials and to reduce the waste.

2.1.1 Circular Economy

It is the extension or the completion of the linear economy model. The Circular Economy (CE) works under the three principles: eliminate waste and pollution, circulate products and items at their peak value and regenerate (Ellen MacArthur Foundation 2015). The increasing popularity of CE has been overwhelming due to its concentration majorly towards resource scarcity and the effects of economic actions on the surroundings (Gupta and Gupta 2019). In this method less material is used, products and materials are made and transported with less cost, the used or waste products are reused as input materials for the new products (McKinsey and Company 2018). The circular economy works under five pillars. They are Sustainable inputs, sharing platform, Product as a service, Life extension, Recycle and Reuse. A tree's life starts from inside the soil and ends inside a soil. A turtle's life starts soil and ends in the soil. That is how Earth was able to survive this long. Similarly, when the CE is integrated into the supply chains it forms a loop thus increasing the efficiency of the product and process (Sarkis and Dou 2017). When old materials are broken down and made into new materials the industry saves money. Large items can also be returned back to the manufacturers at the products end life, so that the used products can have a new life (Bekrar et al. 2021). If every used electronic item is recycled or reused by the manufacturers then tons of waste can be reduced and raw materials consumption will be reduced. Manufacturers can also use the resources effectively. This in turn increases the survival of every living being on earth.

In Fig. 2.1, the grey lines represent the circular flow of the renewables whereas the blue lines represent the circular flow of the finite materials.

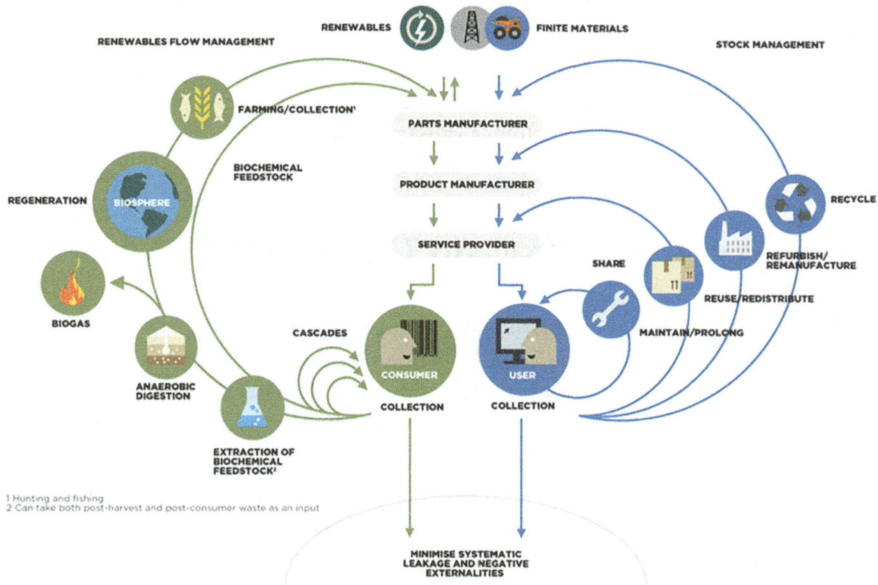

Fig. 2.1 Circular economy systems. (**Source**: Ellen MacArthur Foundation. *Circular economy systems diagram* (February 2019). www.ellenmacarthurfoundation.org. Drawing based on Braungart & McDonough, Cradle to Cradle (C2C) (Ellen MacArthur Foundation 2015))

2.2 Integration of Circular Economy into Supply Chains

Circular Economy is an economic growth model with an effective utilization and recycling of resources retaining its essential characteristics for development (Luthra et al. 2022). Integrating circular economy into supply chains is paramount as it helps businesses to reduce its long-term costs, prevents accumulation of rubbles, and increases network value and growth opportunities to businesses. CE also makes sure that life of raw materials is extended as long as it can with prime quality (Chopra et al. 2019). CE could increase jobs for thousands of people and reduce carbon emissions around millions of tons. In order to integrate CE into Supply chain people and socio-eco systems must coherent together (Achillas et al. 2018; Krestyanpol 2022). Industrial symbiosis plays a vital role in exchanging the by-products, energy and materials between the producers. Suppliers play a key role in safekeeping the environment. Increasing the Industrial symbiosis will create a bond between the suppliers and the manufacturers. Fewer trips with small travel distance will make the bonding process easier as it cuts down the costs and affects the environment less. The market scenario is directly affected by the customers' behaviour. The customer also needs to change his views on using his products and his impact on the environment if CE is to be integrated effectively (Ferguson and Souza 2010). Government should create remanufacturing policies, awareness about CE. The fundamental principle in applying the CE is the reverse logistics. One of the concepts that is widely used in closed loop supply chain management is Remanufacturing (World Economic Forum 2017). Even if it is the most widely used concept it is effective only if the produced products have the ability to be remanufactured. Many companies who have implemented the circular supply chain have witnessed the increase of their profit and in customer satisfaction.

2.3 Ways to Achieve Circularity Through Supply Chain

While it is believed that a circular supply chain will better recover value from end-of-life products to produce zero waste, value chain collaboration is viewed as a crucial facilitator for the process of transition, and the characteristics needed to put this collaboration into effect are still to be found. Closing the loop in a circular supply chain means that products are designed to be reused, repaired, or recycled at the end of their useful life. This can be done through a variety of methods, such as share, maintain or prolong, reuse or redistribute, recycle.

2.3.1 Share

Sharing platform is one of the circular business models where in collaborations are made among the customers to maximize the value and usage of a product. There are many products like chairs, cars in our home that are unused for a long period of time and they lose their value by being idle. Sharing these unused valued products will help them to have a useful lifecycle. Despite many benefits, sharing such information is uncommon in current supply chains (Akbarieh et al. 2020). By sharing the ownership, the raw materials consumed by the manufacturers gets reduced. To make the economy more circular, information sharing is considered a key enabler (Wijewickrama et al. 2021b). Information sharing benefits certain type of customers while negatively impacts others, the benefiting customers should offer incentives to the lateral (Dominguez et al. 2020). In cases like these, the governments are required to share information from the companies regarding the CE (Huang et al. 2019). Figure 2.2 represents few applications where sharing the products can be effective.

2.3.2 Maintain or Prolong

The main focus of circular economy is to reduce landfills and increase the life of materials and products. Maintenance plays a major role in obtaining high circularity levels. A high life cycle product has partial functional failure. This product requires maintenance than repaired. Maintenance plays a large role in remanufacturing and in repurposing too. Maintenance plays a vital role in achieving circular economy in

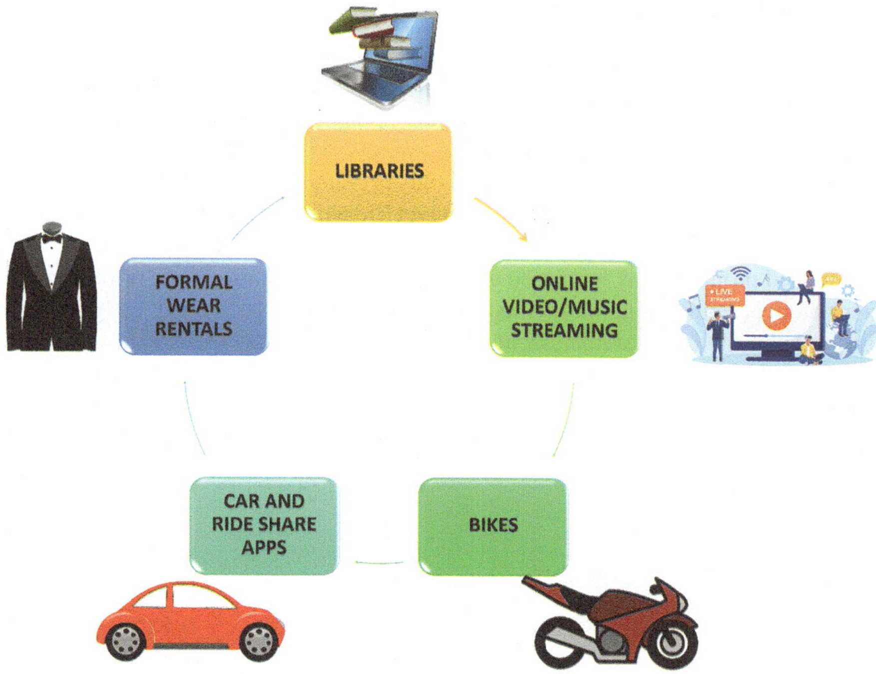

Fig. 2.2 Sharing of products or resources

both within and outside the industry (Khan 2020). It is also required throughout the life of a product or a component to increase its life span. Maintenance also reduces the requirement of raw materials by increasing the life time of products and components of users that are in use. Preventive maintenance in industries and manufacturing units will also increase the life span of machines, tools and products. Profit is achieved can be achieved by increasing the lifespan of a product or a material (Muhammad et al. 2021). In order to reuse a product, maintenance plays a hidden role. For products to be resold or reused they have to be repaired and maintained. The new owner will need to maintain the used product if it is not serviced of maintained properly by the previous user. One of the main activities in refurbish and repair process is the maintenance. To execute the circular strategies the customers are offered with maintenance for a fee. Sometimes they are also offered for free. In this view, maintenance is a form of business model than a strategy. Preventive maintenance is required for the extension of life for a product.

2.3.3 *Reuse or Redistribute*

The word reuse or redistribute is used when a product is used again and again until it reaches its end life. When a good performing product is distributed, it may or may not return back to the previous user but its life cycle increases. 'Reuse' is used for an object without any reprocessing which only prolongs the life of a product (Sinhaa and Modak 2021). A product is given a second life when it is resold or reused, therefore materials required to be produced for the same purpose is less (Weetman 2016). The reused products may be reused as the same product as the whole or its components alone. Requirement of new materials to make a new product is greatly reduced thus reduces the costs to manufacture a new product. Reuse or redistribution also decreases the negative impact on environment. Reuse: Products are returned to the manufacturer or retailer for reuse. Almost half of our used products like automobile, papers and mobile phones can be reused. By reusing a product sales cost and production cost gets reduced.

2.3.4 *Refurbish or Remanufacture*

Refurbishing is another important process in circular economy, which aids the well-being of the environment. In this process products are disassembled and the components are reused to create new products. The products are returned to the manufacturers through vendors or third parties. Remanufacturing reduces the production cost thus reducing the sales cost for the remanufactured product. It also benefits the society by increasing the usage time of end of use items (Wijewickrama et al. 2021a). Refurbished products will have the same quality as the new products. Used products undergo processes like disassembly, sorting, cleaning and testing before it enters the market. Carbon footprint and material consumption are greatly reduced. Energy consumption of the remanufactured products is also less compared to the newly created products. An industry can remanufacture 7–11 more products with an equal number of raw materials and energy consumed. Moreover, remanufacturing creates employment opportunities to both inexperienced non-skilled workers and experienced skilled workers. Germany and France are one of the largest remanufacturers in Europe (Singhal 2020).

2.3.5 *Recycle*

Recycling is the process of gathering and processing components or materials after it has been used and thrown away as trash. Products are broken down into their raw materials and used to create new products. Most of the time, the broken parts are made into a completely new product than a newer version of its previous form.

Recycling is the process of gathering and processing components or materials after it has been used and thrown away as trash. Recycling rare earth metals plays a significant role in reverse logistics (Mishra et al. 2012). Manufacturers repurpose raw materials like plastic, aluminum, cardboard, paper, metals, and glass in a circular supply chain. They restore and resell items that were previously owned. Rather than selling their products, they and the retailers rent them. Additionally, they select eco-friendly packaging and storage options like recycled pallets. For example, paper cups, plates, newspapers are shredded and processed to make a new product like paper box, notebooks and similar products. Sweden has the most recycling rate of 99% which is the best in the world. They are almost running out of trash. Germany is next in line with a recycling rate of 66.1% followed by Austria which has 54%. Recycling can benefit the environment and community. If a product cannot be reused or reduced, it then can be recycled.

Closing the loop in a circular supply chain has a number of benefits, including:

2.3.5.1 Reduced Waste

By keeping products and materials in use, circular supply chains help to reduce waste. For example, if we keep recycling plastics and other products there won't be any need to dispose.

2.3.5.2 Conservation of Resources

Circular supply chains help to conserve resources by reducing the need to extract new materials.

2.3.5.3 Improved Environmental Performance

Circular supply chains can help to improve environmental performance by reducing emissions and pollution. Earth has only 1% of safe pollution level.

2.3.5.4 Economic Benefits

Circular supply chains can create new jobs and opportunities in the recycling and reuse industries. Developing countries will benefit more if they adopt Circular supply chain in their country.

2.4 Challenges of Circular Supply Chains

There are a number of challenges to implementing circular supply chains, including:

2.4.1 Cost

It can be more expensive to design and manufacture products that are recyclable or reusable. Key notes like consuming area geographical location are important to implement the remanufacturing concept and to cut down the cost of a product. The suppliers must create an efficient supply loop to reduce the transportation cost. The reduced transportation cost will help to reduce the manufacturing cost. The reduced manufacturing cost will reduce the product cost which will encourage the customers to buy a reusable or recyclable product. Another cost to note is the cost of the used products to return to its manufacturers. Businesses whose expansion is directly dependent on limited resources will be at a disadvantage in the marketplace. This is a result of prices that are increasing and fluctuating, which makes it harder for them to predict and compete with rivals that require fewer resources. This can be reduced if circular supply chain is properly designed.

2.4.2 Consumer Behaviour

Consumer behaviour plays a vital role in the long-term success of businesses that intertwine circular initiatives. Consumers may not be willing to pay more for products that are designed for a circular system. An awareness and education programmes should be created among the people regarding the Circular supply chain management. Consumer behaviour is a group of action and reactions of social in the field of consumption (Atabaki et al. 2020). Consumer's behaviour towards their consumption of the products and the amount of waste they produce also must change. Sometimes traditions and elder's opinion does not allow the current generation to adapt to the changes. A collaborative model consisting of the consumer, supplier and retailer will help the consumer to adopt or trust the Circular supply chain. The Circular supply chain must be designed in such a way that it motivates the consumers, manufacturers and suppliers to adopt them thus by taking a green initiative. CE involves a change in customer behaviour, industries and the economy through shifts in rules, behaviours in society (Chizaryfard et al. 2021).

There is a lack of clear regulations and standards for circular supply chains. Regulations related to SCM are vague and hard to understand by the manufacturers and their suppliers. Governments should insist the manufacturers to produce products that can be reused or recycled. There are very few schemes on implementing the Circular supply chain and it must be increased. The policies on certain goods must be changed so that the manufacturers can design a new process according the customer's requirements. Governments should also increase the demand for sustainable products which is less now. With the raise in demand and the awareness of the consumers in using the sustainable products circular supply chain management will function with effectiveness includes economic interest and social interaction.

2.5 Opportunities of Circular Supply Chains

Logistics is the driving force behind the circular economy, hence its business sectors are evolving or growing, and perhaps new ones are being created. This also holds true for points of contact and consumer interactions. Recirculation opens up new marketplaces that can be used to reach new client segments, such as those for pay-per-use models or spare parts. From being service suppliers to being valuable contributors to the supply chain, logisticians can develop. There are noticeably fewer international procurement transports since the circular economy lowers the need for basic raw materials, even while products are still in demand. In addition, there is a growing need for both regional and local transportation, particularly as a result of the movement of products. One product will split into numerous distinct items, each of which will require unique logistical solutions tailored to its own conditions; bundling ideas will also become crucial. Service providers need to provide offers that are both socially acceptable and sustainable, meaning they won't worsen the environment's carbon footprint.

New business sectors pertaining to logistics are also developing concurrently. Returned items, for instance, have to be separately dismantled, cleaned, or fixed before being put back on the used market to be used as raw materials or replacement parts. Logistics service companies are already well-positioned to handle these additional responsibilities in addition to transportation and storage. Compared to typical business, these services with value also have better profit margins. The automotive sector can act as an example. There, pre-assembly of components is one of the long-established services.

2.6 Conclusion

Building a strong foundation requires a collaborative action. Moving to circular economy will influence all sectors and government policies. Circular supply chains offer a number of benefits for businesses, consumers, and the environment. As the demand for sustainable products grows and government support increases, the opportunities for circular supply chains will continue to grow. Circular supply chain produces good yield whilst consuming less resources. The emission rate also decreases for the reusable and cycle able products. Less amount of waste is added to the environment by Circular supply chain. Not only the government but also the consumers must take necessary steps in implementing the circular supply chain. Landfill waste causes about 11% of earth's methane emissions. It is predicted to rise up to 70% by 2050 and air pollution has affected 99.82% of earth with levels of PM2.5. Circular supply chain takes time and money to implement but has greater good compared to linear supply chain. It is necessary that every organization must adapt to circularity in their supply chain to have a sustainable future.

References

Achillas C, Bochtis DD, Aidonis D, Folinas D (2018) Green supply chain management. Routledge, New York

Akbarieh A, Jayasinghe LB, Waldmann D, Teferle FN (2020) BIM-based end-of-lifecycle decision making and digital deconstruction: literature review. Sustain For 12:2670

Atabaki MS, Mohammadi M, Naderi B (2020) New robust optimization models for closed-loop supply chain of durable products: towards a circular economy. Comput Ind Eng 146:106520

Bekrar A, Ait El Cadi A, Todosijevic R, Sarkis J (2021) Digitalizing the closing-of-the-loop for supply chains: a transportation perspective. Sustain For 13:2895

Chizaryfard A, Trucco P, Nuur C (2021) The transformation to a circular economy: framing an evolutionary view. J Evol Econ 31(2):475–504

Chopra S, Meindl P, Kalra DV (2019) Supply chain management. Pearson

Dominguez R, Cannella S, Ponte B, Framinan JM (2020) On the dynamics of closedloop supply chains under remanufacturing lead time variability. Omega (U.K.) 97:102106

Ellen MacArthur Foundation (2015) Towards the circular economy: Opportunities for the consumer goods sector. Ellen MacArthur Foundation, pp 1–112

Ferguson ME, Souza GC (2010) Closed-loop supply chains – new developments to improve the sustainability of business practices. CRC Press, Boca Raton

Gupta S, Gupta A (2019) Dealing with noise problem in machine learning data-sets: a systematic review. Procedia Comp Sci 161(2019):466–474

Huang H, Meng Q, Xu H, Zhou Y (2019) Cost information sharing under competition in remanufacturing. Int J Prod Res 57:6579–6592

Khan SAR (2020) The critical success factors of green supply chain Management in Emerging Economies. Routledge

Krestyanpol L (2022) Simulation modeling of consumer behaviour within the concept of smart consumption. Procedia Comp Sci 217(2023):774–783

Luthra S, Kumar A, Sharma M, Garza-Reyes JA, Kumar V (2022) An analysis of operational behavioural factors and circular economy practices in SMEs. J Bus Res 141:321–336

McKinsey & Company (2018) The business case for a circular economy

Mishra N, Kumar V, Chan FTS (2012) A multi-agent architecture for reverse logistics in a green supply chain. Int J Prod Res 50(9):2396–2406

Muhammad RA, Tjahjono B, Ridlo KD (2021) A decision process for the applications of artificial intelligence in sustainable operations and supply chain management. Energy Policy 1–12:1487

Orji IJ, Ojadi F (2023) The circular supply chain – basic principles and techniques. CRC Press, Boca Raton

Sarkis J, Dou Y (2017) Green supply chain management: a concise introduction. Routledge, New York

Singhal D, Tripathy S, Jena SK (2020) Remanufacturing for the circular economy: Study and evaluation of critical factors. Resources, Conservation and Recycling 156:104681

Sinhaa S, Modak NM (2021) A systematic review in recycling/reusing/re-manufacturing supply chain research: a tertiary study. Int J Syst Eng 14:1411–1432

Weetman C (2016) A circular economy handbook for business and supply chains: repair, remake, redesign, rethink. Kogan Page, London
Wijewickrama MKCS, Chileshe N, Rameezdeen R, Ochoa JJ (2021a) Information sharing in reverse logistics supply chain of demolition waste: a systematic literature review. J Clean Prod 280:124359
Wijewickrama MKCS, Rameezdeen R, Chileshe N (2021b) Information brokerage for circular economy in the construction industry: a systematic literature review. J Clean Prod 313:127938
World Economic Forum (2017) The circular economy: a powerful way to decouple economic growth from environmental degradation

Chapter 3
Transitioning of Traditional Supply Chain to Circular Supply Chain

Abstract Applying the concept of the circular economy to the supply chain means moving from the linear models of a traditional supply chain to a new model of a circular supply chain that is entirely different in terms of ideas about resources and sustainability and economic stability. Most resource supply chains have relied on the successive linear value system that can be described as take-make-dispose process which suffers from resource over-exploitation and waste production. On the other hand, circular supply chains are those that try to reutilize materials and components as many times as possible in various forms in order to reduce wastage thereby embracing practices like recycling, remanufacturing as well as reusing products and the prolonging of their useful life. This chapter analyses the factors that facilitate organisations to pursue circular supply chain management as well as the key barriers: innovation in product design, supply chain management, and customer behaviour shift. It also provides examples of organisations that have established circular principles and witnessed that technology and data are core to circular journeys. Lastly, the purpose of this research is to deconstruct the multi-faceted transition towards the circular supply chain, so that the stakeholders have a clear picture of how the journey will be made, and in the process, it may offer a chance at better sustainability, improved economic returns, and thus, a better hold on the competitive base within the global market.

Keywords Circular economy · Supply chain management · Sustainability · Recycling · Remanufacturing · Product design

3.1 Introduction

Transitioning a supply chain from a traditional linear model to a circular model fundamentally alters how businesses manage resources throughout their lifecycle. The traditional linear model typically follows a 'take-make-dispose' approach, which is increasingly recognized as unsustainable due to the resulting waste and resource depletion. In a circular supply chain, the focus shifts to resource efficiency

J. Kandasamy et al., *Circular Supply Chain Management*, Circular Economy and Sustainability, https://doi.org/10.1007/978-3-031-81750-2_3

and sustainability (Afteni et al. 2021). Instead of discarding materials after one use, businesses aim to keep resources in use for as long as possible through various strategies such as recycling, refurbishing, and remanufacturing. This requires businesses to rethink their sourcing strategies, opting for renewable or recycled materials whenever possible. Minimizing waste is a primary goal of transitioning to a circular model. Companies must implement processes that reduce waste generation at every stage, from production to consumption. This entails designing products that are easier to disassemble and recycle, allowing for a more seamless integration back into the production cycle. Maximizing the reuse of materials involves creating systems for product return and refurbishment. Effective return systems are critical; they ensure that end-of-life products are re-entered into the supply chain for further use rather than being discarded. Companies may adopt strategies such as buy-back programs, leasing models, or customer incentives to encourage the return of used products. Ultimately, this transition contributes to a more sustainable business model that benefits not only the company but also the environment and society as a whole. By reducing reliance on raw materials and minimizing waste, businesses can significantly lower their environmental footprint. This is crucial in addressing pressing issues such as climate change and resource depletion. Moreover, a commitment to sustainability can enhance a company's brand image, aligning it with the values of eco-conscious consumers (Haar 2024). Customers want for the instant satisfaction that comes with quick shipment. Modern solutions are replacing outdated techniques of supply chain management since they can't keep up. Retailers are experiencing supply chain issues due to declining efficiency and increasing difficulty in getting goods where they need to be on time. So how can companies get beyond these obstacles and adjust to a market that is changing quickly? via the application of creative management techniques. Three components make up a supply chain, and each one serves a specific purpose to ensure the smooth operation of the whole system. The supply chain includes every step of the product's journey, from the supplier to the manufacturer to the client, from raw materials to the final product delivery to your customer. The entire system is affected if any one component of the process is interrupted or delayed.

3.2 Conventional Supply Chain

Designing, organizing, executing, and regulating supply chain processes with the goal of generating net value and creating a competitive infrastructure is known as traditional supply chain management. In doing so, it is ensured that global logistics services are used, supply and demand are synchronized, and performance is measured worldwide (Michelini et al. 2017). Coordination and cooperation with channel partners, which might include suppliers, middlemen, outside service providers, and customers, are essential to managing a traditional supply chain. To assist businesses compete in the marketplace, traditional supply chain management essentially combines the management of supply and demand both inside and between organizations (Fig. 3.1).

The industrial "take, make, dispose" cycle and the corresponding lifestyles that rely on it deplete finite resources to produce goods with limited shelf lives that wind up in incinerators or landfills. In contrast, the cyclical method draws lessons from biological systems. The phrases "closed loop" and "regenerative" are typically connected with it because it believes that our systems should function like living things, digesting nutrients that can be fed back into the cycle, whether it be biological or technological. Numerous schools of thought might claim or use the general term "circular economy," yet they all converge on the same fundamental ideas (Shaikh et al. 2024).

The Linear supply chain handles company and customer orders by going through a few fundamental steps. Gathering of base materials at this phase, all of the raw materials needed to manufacture and produce the finished product are gathered; these raw materials might consist of a single ingredient or several elements. The fundamental steps of the conventional supply chain include the stage of material collection from vendors. Which, via the proper suppliers, allows the production or manufacturing process to receive the necessary raw materials. The goods are produced following the identification of the raw materials needed for the workday and their procurement from vendors. Several pieces of equipment may be employed in this process to complete the items and produce a finished good that can be given directly to clients. Followed by supplying to the consumers this is a component of the conventional supply chain, which is used to deliver manufactured goods to consumers or merchants. The final step of the process where merchants or end users buy the completed goods directly from the manufacturer. Effective supply chain management may reduce total costs and increase profitability for a business (Bhattacharya et al. 2024). A broken link might have expensive consequences for the entire network. There are several supply chain model options. The model that a business chooses will rely on its unique demands and organizational structure. Here are few instances:

Continuous Flow strategy: Businesses that manufacture identical goods with little fluctuation find success with this conventional supply chain strategy. There should be a lot of demand for the items and not much need for redesign. Because there is less variation, managers are able to optimize production schedules and maintain strict inventory management. Managers must routinely restock raw supplies in a continuous flow paradigm to avoid production bottlenecks.

Fast Chain Model: This model functions best for businesses who market their goods in accordance with current fashions. Companies that employ this strategy must

Fig. 3.1 Linear supply chain

launch their goods swiftly in order to capitalize on the current trend. From concept to prototype to manufacturing to customer, they must move quickly. One sector that makes advantage of this supply chain paradigm is fast fashion.

Adaptable Model: Businesses that produce items for the holidays or seasons frequently employ the flexible approach. These businesses have spikes in demand for their goods, which are followed by protracted dips in or absence of demand. Their ability to swiftly ramp up for production and effectively shut down when demand wanes is ensured by their flexible model. They have to be precise in estimating their labour, inventory, and raw material needs if they want to turn a profit.

3.3 The Best Practices in Supply Chain Management

3.3.1 Ongoing Development

Suppliers' management systems in successful supply chains are always a work in progress, constantly evolving as the market and the needs of the businesses evolve. Another continuous learning includes periodic customer training with the objective of acquainting them with the current trend, approaches or tools in the industry. Knowing supply chain environments are constantly evolving, companies must endeavour to build learning culture within the staff so that the latter can be ready to face a widening array of emerging issues. Hiring the right talents becomes a key enabler through which organisations increase the efficiency and innovation of their supply chain (Tseng et al. 2024).

3.3.2 Higher Speed

The aspect of speed is central to supply chain management since customer expectations are only measurable by how fast the whole process is accomplished. The role of supply chain management covers making changes in supply chain networks to increase efficiency and seek for avenues which, when optimized, would minimize on lead times in product delivery. In logistics, organizations can employ just-in-time inventory control so that they can manage their stocks effectively thus responding to the market needs. A faster response may also result to better services, increased satisfaction and hence, increased rate of customers' retention (Ferreira et al. 2024).

3.3.3 Promoting Cooperation

To be efficient, interactions between different records of the chain of supply are fundamental. Effective systems develop collective relations between suppliers, manufacturers and distributors in the accomplishment of organizational goals. Some of the areas that should be communicated freely, solved together and many more are; Ineffective communication leads to miscommunication of information and work. Firms that advocate collaboration not only build confidence and credibility but also are in a better position to address forces such as disruption of the supply chain, or when there is a change in demand (Berardi and Peregrino de Brito 2021).

3.3.4 Leveraging New Technology

Applying new technologies in the supply chain management is important for improving flow and performance in the processes. Best practices for the supply chain management systems are always on the lookout for better technologies to work with, including automation and artificial intelligence, big data analytics, and others (Lu et al. 2024). Real time technologies can make this a reality and help the supply chain managers in their endeavours of trend spotting, demand forecasting and making good decisions. Hence, through innovation, companies are able to cut costs, increase speed hence achieving an optimal supply chain.

3.3.5 Implementing Metrics

For evaluation and to highlight, gaps in the supply chain, supply chain management systems set up performance goals or Key Performance Indicators (KPIs) and Metrics. These metrics enable the staff members to assess resultantly each stage of the supply chain, some of the factors that may be measured range from one or more of the following; stock turnovers, order accuracy and delivery time. Through such measurements, executives and managers gain the necessary knowledge of supply chain susceptibilities and can make changes that encourage enhancement of the system. Performance evaluations based on such measures also make it possible to check that the supply chain strategies are in tune with the overall organizational goals, and that there is the flexibility to deal with any changes.

3.4 Circular Supply Chain's Goals

A closed-loop system where goods, materials, and resources are continually recycled, repurposed, and replenished is known as a circular supply chain. A circular supply chain seeks to maximize value creation over the long term, decrease waste, and lessen its negative effects on the environment (Amir et al. 2023). This necessitates a holistic approach that takes into account the full life cycle of materials and products, from design to disposal at the end of their useful lives as shown in Fig. 3.2.

One approach to resource management that puts sustainability and environmental preservation first is the circular supply chain concept (Manu et al. 2022). Under the circular model, items are made with the intention of being reused or repurposed at the end of their existence, in contrast to the conventional linear supply chains where products are created, utilized, and discarded as trash. The idea behind this strategy is straightforward: firms should restore broken or outdated items for future use rather than disposing of them in landfills. Maintaining resources in closed loops, where they continue to fulfil a purpose, helps to minimize consumption and eliminate waste. In order to practice circular supply chain management, products must be designed with ease of repair, upgradability, and modularity in mind, as well as energy efficiency and environmentally friendly materials. Businesses may lessen their carbon impact and extend the life of their products by factoring these steps in during the production planning stages. Companies at all levels, from producers to consumers, must work together to implement a circular supply chain model. It's a creative approach to resource management that encourages sustainable business practices and contributes to environmental protection.

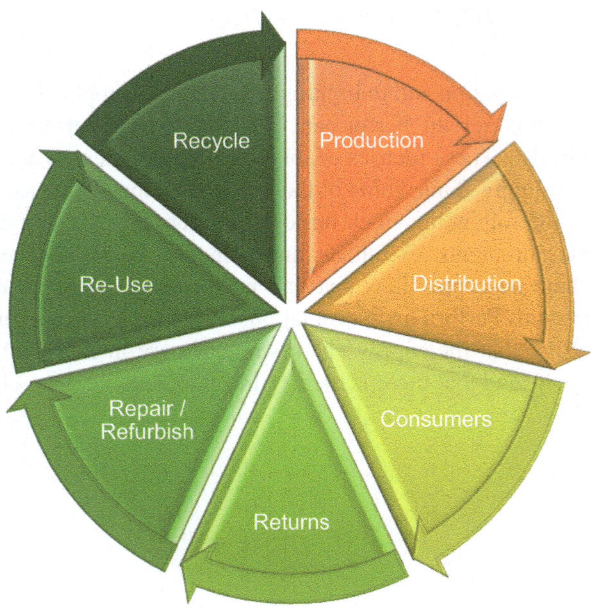

Fig. 3.2 Circular supply chain

3.5 Circular Supply Chain (CSM) Approach

A novel approach to supply chain management, the Circular Supply Chain Model concentrates on developing environmentally friendly and sustainable business practices. Reuse, repurposing, refurbishing, recycling, returns, and distribution are the guiding concepts of the approach. In order to put the circular supply chain concept into practice, items must first be designed with sustainability in mind (Amir et al. 2023). This implies that businesses should minimize their environmental effect throughout the production process and employ recyclable or biodegradable materials. Subsequently, goods are delivered straight to consumers or retail establishments via routes designed to reduce waste by using less packing materials.

A product can be repaired or reused for a different use after it has reached the end of its life cycle for its original function rather of being thrown away. Recycling makes sure that priceless resources are recognized and used again in other ways when this is no longer feasible. Additionally, consumer returns can be gathered and handled again. Businesses may assess production issues and reduce mistakes in the future by gathering returned goods from customers at one place. The circular supply chain model reduces waste at every level of the supply chain process and prioritizes long-term sustainability over immediate financial gain. Although putting the Circular Supply Chain Model into practice in your company might be difficult, the effort is worthwhile. Finding places where you can cut waste and boost efficiency is the first step. This might entail recycling resources, repairing them for resale, or reusing or repurposing them. For this model to be properly implemented, you must set up specific objectives and metrics to track your progress. Employee participation in waste reduction initiatives and education on the value of sustainability are also essential (Bai et al. 2024).

For a circular supply chain to be implemented, you must engage with suppliers that share your commitment to sustainability. This might entail collaborating with vendors who use sustainable sourcing methods or sell reconditioned goods (Berardi and Peregrino de Brito 2021). Regularly assessing the results of these efforts and making necessary modifications are crucial. You may save expenses, boost customer happiness, and build a more sustainable future for your company by continuously enhancing your circular supply chain model. All parties engaged in the process must continue to put up work and attention in order to implement a circular supply chain. On the other hand, by moving toward a more sustainable strategy, firms might eventually reap financial and environmental rewards.

3.5.1 *Six "R's": The Success of Circular Supply Chains (Khan et al. 2024)*

Rethink/Re-imagine: In order to get more out of your supply chain, how can your business rethink its approaches to material procurement, production, packaging, and delivery procedures, as well as product and material End-of-Life strategies? It might be useful to analyze industry and rising market trends to find areas that need to be reconsidered.

Refuse—Seek ways to avoid needless waste by declining to utilize inefficient materials such as single-use plastic and specific kinds of glass, chemicals, and ceramics.

Reduce—Assess your production procedure to find ways to cut back on the amount of resources used and the need for additional supplies. Putting overproduction reduction measures into practice is also essential to creating a circular supply chain that is sustainable. Since overproduction is the primary source of waste in the industrial sector, any mitigation measures can have a significant negative influence on the environment as well as revenue.

Reuse—To reduce waste and increase utility, find ways to repurpose resources, faulty items, and consumer returns. This greatly lowers your landfill contributions while reducing lost time, effort, and money.

Repurpose—It's time to use your creativity! How can you increase your efficiency by repurposing goods or resources that might otherwise go to waste? Once you get an answer to that query, you may begin cutting costs and your environmental effect.

Recover—A crucial component of the circular economy in general and the supply chain in particular is recovery. In the linear supply chain business model, this "R" phase takes the role of the "Dispose" step. It seeks to optimize the value and utility of precious resources and goods by reintroducing them into your supply chain (Fig. 3.3).

3.6 Circular Supply Chains vs Linear Supply Chains

Circular supply chains and linear supply chains are two fundamentally different approaches to managing resources and materials in the supply chain as shown in Fig. 3.4.

In a linear supply chain, resources are extracted, processed, manufactured, utilized, and eventually disposed of as trash. This is a classic, linear approach to supply chain management. This method is sometimes called a "take-make-dispose" model. Long-term sustainability is impossible for linear supply chains due to their high resource consumption, significant waste production, and pollution. On the other hand, a circular supply chain is a supply chain management strategy that seeks to

Fig. 3.3 6R's

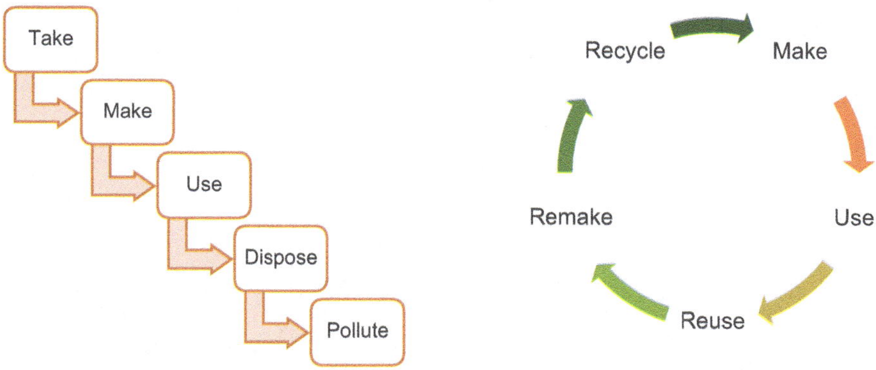

Fig. 3.4 Linear versus circular

maximize the longevity of goods and resources (De Angelis et al. 2018). This is accomplished by developing closed-loop systems that facilitate material recovery and reuse as well as by designing goods and procedures for reuse, recycling, and regeneration. The goal of circular supply networks is to cut down on waste and the consumption of virgin resources.

In a circular supply chain, resources are retained in use for as long as feasible, whereas in a linear supply chain, resources are taken and consumed in a one-way process. Whereas circular supply chains emphasize material recovery and reuse as well as trash minimization, linear supply chains produce a lot of waste and pollution. Linear supply networks frequently give priority to performance and cost in product design, circular supply chains demand that end-of-life situations be considered. Linear supply networks frequently function in a compartmentalized fashion, circular supply chains necessitate cooperation amongst supply chain stakeholders to develop closed-loop systems (Nasir et al. 2017). There is a difficulty to change in linear supply chain, whereas circular supply chains depend on innovation in business models, procedures, and technology to achieve circularity.

3.7 Stages During Supply Chain Transition Towards Circular Supply Chain

Lessons learned from the round table of span industries cover a wide variety of topics, such as adoption levels and obstacles to collaboration, in addition to industries and waste streams/end-of-life items. But a crucial lesson is that adopters must cultivate circular ways of thinking if they want to transition to circular supply chains as shown in Fig. 3.5 (Taddei et al. 2024). A mindset like that contains a lot of components (Afteni et al. 2021). These three round table examples are crucial in helping people adopt a circularity-focused mindset.

Fig. 3.5 Thoughts of CSM

3.7.1 *Alter Ingrained Beliefs*

Thinking in circles means challenging established paradigms, and when required, rethinking and reformulating them. Consider a material or product that has reached the end of its useful life instead of viewing it as garbage and instead seeing it as the beginning of a new process that will extract its worth. One participant provided a concise summary of the idea when they said that "waste is just excess material in the wrong person's hands." Throughout the day, this illustration of a novel viewpoint spurred conversations at the round table and encouraged participants to reconsider waste and end-of-life goods. It is feasible to envision alternative methods of using products that do not depend on the supply of virgin materials and conventional disposal techniques, provided that the appropriate take-back supporting structures are in place. Additionally, even if there are many chances to advance on the path to circularity, businesses must have an open mind while pursuing circular methods of conducting business. Redesigning things, prolonging their life cycles, cutting waste, and developing whole new business models are among the choices (Roy et al. 2022). Several businesses at the round table have tested prototype product lines using recycled and post-consumer recycled materials. They have learned what approaches work and how to effectively convey their findings by using the expertise they have gathered.

3.7.2 *Utilize Internal Buy-in*

The desire for the supply chain to become more circular is not necessarily driven by consumers; demand to change may also originate from inside the company. During the round table, a prominent sports consumer goods firm said that their initiative to recycle end-of-life items was initiated by staff members who were worried about the amount of garbage the company was producing. Employees redesigned a procedure to dismantle old goods and use the materials that might be recovered.

3.7.3 *External Partnerships to Generate Shared Benefit*

Businesses shouldn't embark on the task of developing a successful circular supply chain alone. Businesses that have made investments in circular supply chains emphasize the value of fundamental alliances with suppliers and other downstream stakeholders. Sometimes forming new alliances is essential to achieving a certain objective. For example, a beverage firm came to the realization that recycling its tea packaging using ways it was accustomed to employing was not feasible. The company collaborated with a fresh supplier to develop a biodegradable packaging option. The substance decomposes into soil suitable for agricultural use.

Considering the complete value chain necessitates a comprehensive strategy when designing a supply chain for a circular economy. This include creating plans to minimize, reuse, and recycle resources as well as mapping the whole supply chain and detecting possible waste streams. To build a closed-loop system, suppliers, customers, and other stakeholders must work together. Thorough analysis of your present supply chain to find areas that might want improvement and possible circularity prospects. This might entail looking at the environmental effect of your supply chain, evaluating the usage of virgin resources, and finding ways to reduce and reuse waste. Create circularity goals that are consistent with the circular economy's tenets. This might entail establishing goals for recycling, resource efficiency, and waste reduction. Establishing goals aids in prioritizing acts that promote circularity and provide a clear course of action. Create items and processes that facilitate simple recycling, reuse, and disassembly. Wherever feasible, try using fewer virgin materials and adding sustainable resources. This might entail rethinking goods to make them more circular and collaborating with suppliers to find sustainable materials. To develop a closed-loop system that considers the full product life cycle, collaborate with suppliers, customers, and other relevant parties. Establishing closed-loop systems, promoting sustainable practices, and identifying possibilities for circularity all need collaboration, which is essential to attaining a circular supply chain. Improve material and product tracking, traceability, and transparency by utilizing technologies like blockchain, artificial intelligence, and the Internet of Things. Through increased efficiency, supply chain visibility, and the development of new circular business models, technology may facilitate circularity. Although implementing a circular supply chain can be difficult and time-consuming, businesses can start laying the groundwork for sustainability and circularity by taking these basic steps. It's not a simple task. It requires a lot of effort and time. However, it is worthwhile!

Working together with all parties involved in the value chain, Collaborate with vendors, clients, and other relevant parties to establish a closed-loop system that considers the complete life cycle of materials and products (Berardi and Peregrino de Brito 2021; Patra et al. 2024). Take end-of-life disposal into account when creating designs that facilitate simple recycling, reuse, and disassembly. To lessen the supply chain's negative environmental effects, choose non-toxic, renewable, and biodegradable materials. To cut down on waste and boost productivity, apply lean manufacturing concepts. Create new business models: Put service-based solutions, circularity, and resource efficiency ahead of product ownership in your new business models. To improve material and product monitoring, traceability, and transparency, integrating technologies like blockchain, artificial intelligence, and Internet of Things. Develop a method for recovering, restoring, and reusing materials and products that have reached the end of their useful lives (Rogan et al. 2022). Construct an environment in which goods, materials, and resources are continually recycled, regenerated, and reused. To monitor the advancement of circularity objectives, set up reporting frameworks and metrics. Raise awareness and instruct staff members on the significance of sustainability and circularity in the supply chain. Educate the consumers of the advantages of circularity and provide them with incentives to take part in the closed-loop system. Choose suppliers who make sustainability and circularity a priority in their business practices. Make sure that all applicable laws, rules, and guidelines pertaining to sustainability and circularity are followed. Use low-carbon transportation modes and optimize logistics to minimize transportation emissions. Use packaging materials that are biodegradable, recyclable, and reusable to encourage circularity and reduce waste. Companies that want to build a more circular and sustainable supply chain that decreases waste, boosts productivity, and generates long-term value must put these best practices into effect. These methods have been shown to be successful in a variety of sectors and can provide fantastic outcomes, particularly when paired with a culture that places a high value on sustainability and lowering carbon footprints (Tiwari et al. 2024). Companies may create good change and a better environment for future generations by implementing circular supply chain strategies and working toward a sustainable future.

3.8 Conclusion

There are several advantages for transiting towards circular supply chain, as it has much to offer the businesses, the environment, and society. By extending the life of resources and goods, a circular supply chain lowers waste. By maximizing transit and utilizing renewable energy, it also reduces emissions. Enhanced resource efficiency, A circular supply chain may enhance resource efficiency and lessen the requirement for virgin resources by preserving materials and reducing waste (Vimal et al. 2019). Improved brand reputation, Businesses may attract environmentally sensitive stakeholders and customers by emphasizing sustainability and circularity in their supply chains (Verleye et al. 2024). By eliminating waste and inefficiencies and generating value from waste streams through recycling and reusing, a circular supply chain may cut costs. To adapt goods and procedures for circularity and create new business possibilities and competitive advantages, a circular supply chain demands creative thinking and innovation. By lowering reliance on finite resources, lowering the risk of resource price changes, and lessening the effects of supply chain interruptions, a circular supply chain can increase resilience. Lowering environmental pollution, generating new jobs in the recycling and repurposing industries, and fostering a more sustainable future for all, circular supply chains may benefit society. There are several advantages for both organizations and the environment with the circular supply chain model. Because it places a strong emphasis on reusing, repurposing, refurbishing, and recycling materials across the supply chain, one of the main benefits is waste reduction. Companies may lower their carbon footprint and save money on raw materials and production expenses by putting circular processes into place. This strategy can also increase consumer satisfaction by offering them more affordable, high-quality reconditioned items. By extending the lifespan of products, circular supply chains also encourage the more economical use of resources. This lessens the requirement for ongoing production and delivery of new products. Moreover, the use of a circular supply chain fosters innovation and cooperation across various businesses. Businesses can collaborate to create novel approaches for repurposing waste materials or to create cutting-edge solutions that advance both society and the economy. There are several advantages to using a circular supply chain model that go beyond environmental sustainability. It encourages the effective use of resources, lower costs, more customer happiness, and a chance for innovation across a variety of industries.

The Circular Supply Chain Model is not a perfect solution, despite its many advantages. It is important to be aware of a few disadvantages before using this approach to your company. One of the difficulties is that it takes a lot of money and effort to establish a circular supply chain. Sophisticated logistical systems must also be in place for businesses to handle the operations of recycling, refurbishment, and returns (Upadhayay and Alqassimi 2018). For tiny enterprises or those with limited resources, this may be costly. One possible disadvantage is that it may need some time and effort to persuade buyers to purchase remanufactured goods. In contrast to new items, buyers could also think they are of inferior quality or untrustworthy. Because of this image, businesses may find it challenging to recoup the expenses of restoring returned items. Furthermore, cooperation and communication amongst various departments and stakeholders in a company's value chain are essential for the effective implementation of a circular supply chain. As such, it could take some time for everyone to completely comprehend their duties and responsibilities. When recycling or repurposing elements from rival goods without authorization or license agreements, intellectual property laws may come up. The circular supply chain model nevertheless has a lot of potential to reduce waste and sustain profitability in the long run, even with these obstacles. This is especially true if businesses intend to invest early in the required technological infrastructure and inform customers about the model's advantages.

References

Afteni C, Păunoiu V, Afteni M (2021) Study on the transition from the linear economy to the circular economy. Annals of "Dunarea de Jos" University of Galati, Fascicle V, Technologies in machine building 39:49–55

Amir S et al (2023) Towards circular economy: a guiding framework for circular supply chain implementation. Bus Strateg Environ 32(6):2684–2701

Bai C, Zhu Q, Sarkis J (2024) Circular economy and circularity supplier selection: a fuzzy group decision approach. Int J Prod Res 62(7):2307–2330

Berardi PC, Peregrino de Brito R (2021) Supply chain collaboration for a circular economy-From transition to continuous improvement. J Clean Prod 328:129511

Bhattacharya A, Srivastava S, Majumdar A (2024) Circular supply chains in manufacturing—Quo vadis? Accomplishments, challenges and future opportunities. Bus Strateg Environ 33:4397–4423

De Angelis R, Howard M, Miemczyk J (2018) Supply chain management and the circular economy: towards the circular supply chain. Prod Plan Control 29(6):425–437

Ferreira JJ, Fernandes AJC, Ramírez-Pasillas M (2024) Start-ups and entrepreneurial ecosystems in the circular economy: a multi-level approach for safe and just planetary boundaries. Int Small Bus J 42(4):416–445

Haar G (2024) Transition to a circular economy. In: The great transition to a green and circular economy: climate nexus and sustainability. Springer Nature, Cham, pp 89–126

Khan S, Haleem A, Khan MI (2024) Enablers to implement circular initiatives in the supply chain: a grey DEMATEL method. Glob Bus Rev 25(1):68–84

Lu H, Zhao G, Liu S (2024) Integrating circular economy and Industry 4.0 for sustainable supply chain management: a dynamic capability view. Prod Plan Control 35(2):170–186

Manu E, Ankrah NA, Bentrar J (2022) Transitioning from a linear to a circular construction supply chain. In: Climate emergency–managing, building, and delivering the sustainable development goals: selected proceedings from the international conference of sustainable ecological engineering design for society (SEEDS) 2020. Springer International Publishing

Michelini G et al (2017) From linear to circular economy: PSS conducting the transition. Procedia Cirp 64:2–6

Nasir MHA et al (2017) Comparing linear and circular supply chains: a case study from the construction industry. Int J Prod Econ 183:443–457

Patra SP, Wankhede VA, Agrawal R (2024) Circular economy practices in supply chain finance: a state-of-the-art review. BIJ 31(7):2197–2216

Rogan J, Fürstenberg F, Wieland A (2022) Shaping the transition from linear to circular supply chains. In: Circular economy supply chains: From chains to systems. Emerald Publishing Limited, Bingley, pp 69–87

Roy T et al (2022) Redesigning traditional linear supply chains into circular supply chains–A study into its challenges. Sustain Prod Consum 31:113–126

Shaikh AR, Qazi AA, Appolloni A (2024) Identification and evaluation of the contextual relationship among barriers to the circular supply chain in the Pakistani context–an interpretive structural modelling approach. Prod Plan Control 35(10):1148–1163

Taddei E et al (2024) Circular supply chains theoretical gaps and practical barriers: a model to support approaching firms in the era of industry 4.0. Comput Ind Eng 190:110049

Tiwari S et al (2024) Role of circular economy, energy transition, environmental policy stringency, and supply chain pressure on CO_2 emissions in emerging economies. Geosci Front 15(3):101682

Tseng M-L et al (2024) Modelling hierarchical circular supply chain management enablers in the seafood processing industry in Vietnam under uncertainties. Int J Log Res Appl 27(1):30–58

Upadhayay S, Alqassimi O (2018) Transition from linear to circular economy. Westcliff Int J Appl Res 2(2):62–74

Verleye K et al (2024) Pushing forward the transition to a circular economy by adopting an actor engagement lens. J Serv Res 27(1):69–88

Vimal KEK et al (2019) Analyzing the drivers of sustainable procurement using the DEMATEL approach. In: Sustainable procurement in supply chain operations. CRC Press, pp 233–245

Chapter 4
Barriers of Circular Supply Chain

Abstract The shift from linear to circular supply chains (CSC) has emerged as a crucial strategy for addressing environmental sustainability challenges. A circular supply chain emphasizes reducing waste, reusing resources, and recycling materials to create a closed-loop system that minimizes environmental impact and conserves resources. Despite its potential benefits, many organizations face significant barriers when adopting CSC practices. These barriers can be classified into internal and external challenges. Internal barriers include lack of expertise, insufficient awareness, resistance to change, financial constraints, and inadequate stakeholder engagement. On the other hand, external barriers stem from governmental regulations, market dynamics, supply chain complexity, and limited collaboration opportunities. Overcoming these obstacles requires a comprehensive approach that includes engaging top management, fostering a culture of innovation, securing financial resources, and establishing effective partnerships with stakeholders. Furthermore, integrating circular design principles and addressing consumer perceptions about the value and quality of circular products are essential steps in making CSC practices more viable. Organizations must also navigate regulatory frameworks and collaborate with external technical support to overcome challenges related to policy, infrastructure, and technological capacity. By addressing these multifaceted barriers, companies can transition successfully to circular supply chains, positioning themselves as leaders in sustainability while benefiting from long-term economic gains.

Keywords Circular supply chain (CSC) · Sustainability · Barriers to adoption · Supply chain integration · Circular economy · Stakeholder collaboration

4.1 Introduction

The Circular Supply Chain (CSC) represents a paradigm shift in how businesses operate, moving away from the traditional linear model that typically follows a 'take-make-dispose' approach. In a linear supply chain, resources are extracted,

J. Kandasamy et al., *Circular Supply Chain Management*, Circular Economy and Sustainability, https://doi.org/10.1007/978-3-031-81750-2_4

used to create products, and eventually discarded as waste at the end of their life cycle. In contrast, the CSC framework seeks to create a closed-loop system where products, materials, and resources are continuously reused, refurbished, and recycled, thereby reducing waste and conserving natural resources (Amiri et al. 2022). The emphasis on reusing, recycling, and remanufacturing within the CSC framework is crucial for minimizing environmental impacts. This approach not only contributes to waste reduction but also enhances the efficiency of resource utilization. By keeping materials in circulation and reducing the need for new raw materials, organizations can limit their carbon footprints and contribute to broader sustainability goals. The shift towards sustainability has become increasingly important as consumers and stakeholders demand greater accountability from businesses regarding environmental impacts.

Adopting CSC practices offers several potential benefits. These include enhanced sustainability, where companies can reduce their ecological impact; cost savings through improved resource efficiency and waste reduction; and improved competitiveness in a market that increasingly values sustainable practices (Ada et al. 2021). The integration of circularity within business operations can also foster innovation by encouraging the development of new business models that focus on service rather than ownership, such as product-as-a-service concepts. Thus, companies that successfully transition to a CSC model can meet consumer demand for environmentally friendly products while positioning themselves as leaders in sustainability. Despite the evident advantages, organizations often encounter significant challenges when striving to implement CSC practices. These obstacles can include technological limitations, lack of adequate infrastructure for recycling and remanufacturing, and the complexity of redesigning processes for materials recovery (Ayati et al. 2022). Additionally, companies may face internal resistance to change, especially if existing systems and processes have become deeply ingrained. The necessity for cooperation and coordination among various stakeholders—including suppliers, customers, and regulatory bodies—adds another layer of complexity to the implementation of CSC strategies. Moreover, the definition and understanding of circular supply chain concepts are still evolving, which can lead to discrepancies in how organizations approach circularity.

The challenges of implementing Circular Supply Chain (CSC) practices arise from multiple sources, primarily encompassing technological constraints, regulatory complexities, and organizational resistance to change. Technological constraints involve the lack of adequate technologies that support circularity, such as systems for waste tracking, resource recovery, and advanced recycling methods (Chen et al. 2024). Regulatory complexities pertain to the often intricate and varying environmental policies that organizations must navigate, impacting compliance and operational practices. Finally, organizational resistance to change typically manifests in the form of established mindsets, fear of disruption, or lack of awareness about circular principles among employees and management. Integrating circular principles into existing supply chains necessitates a fundamental shift in management practices and strategies. This transformation involves rethinking the traditional linear supply chain model, which relies on a linear flow of resources, to a more circular model that emphasizes reuse, recycling, and remanufacturing. As companies adopt circular practices, they need to develop new processes that enable product take-back, refurbishment, and greater product lifespans, which often requires significant changes in how products are designed and produced (Farooque et al. 2019). This shift demands not only new technologies but also a re-evaluation of supplier relationships and material sourcing strategies, emphasizing collaboration across various stakeholders.

Accompanying the shift in supply chain management is a need for new skills and partnerships that support circular operations. Employees may require training in sustainability practices, systems thinking, and innovative design to effectively contribute to circular initiatives. Furthermore, building partnerships with recycling firms, refurbishment specialists, and even competitors can facilitate knowledge sharing and resource pooling, which are essential for creating effective circular supply chains (Luthra et al. 2022). Establishing these collaborative relationships can prove challenging, especially in industries traditionally characterized by competition rather than cooperation. Many organizations encounter significant deficiencies in the necessary infrastructure and support systems for transitioning toward circularity. This includes physical infrastructure, such as facilities for recycling and remanufacturing, as well as digital solutions that ensure transparency and efficiency throughout the supply chain. The absence of a robust waste collection and management system can further complicate the logistics involved in capturing and processing end-of-life products. As many companies operate under the constraints of existing linear supply chain systems, retrofitting these operations to incorporate circular elements can be daunting and resource-intensive (Mangla et al. 2018). The combination of these factors results in a complex landscape that organizations must navigate to implement effective CSC practices. The interplay between technological, regulatory, and organizational challenges creates a multi-dimensional hurdle that requires comprehensive strategies and careful planning. Effective transition to circular supply chains necessitates a long-term commitment and willingness to invest in research, innovation, and workforce development to overcome these barriers. Thus, organizations that proactively seek to address these challenges are better positioned to benefit from the sustainability and economic advantages that circular supply chains offer.

4.2 Internal Barriers

Internal barriers refer to challenges that originate within the organization itself. These may include a lack of awareness or understanding of circular economy principles among employees and management. Resistance to change and traditional linear business models can hinder the adoption of circular practices. Insufficient investment in research and development for eco-friendly technologies or reluctance to redesign products for recyclability may also be internal barriers (Govindan and Hasanagic 2018; Mehmood et al. 2021). Additionally, a lack of collaboration and

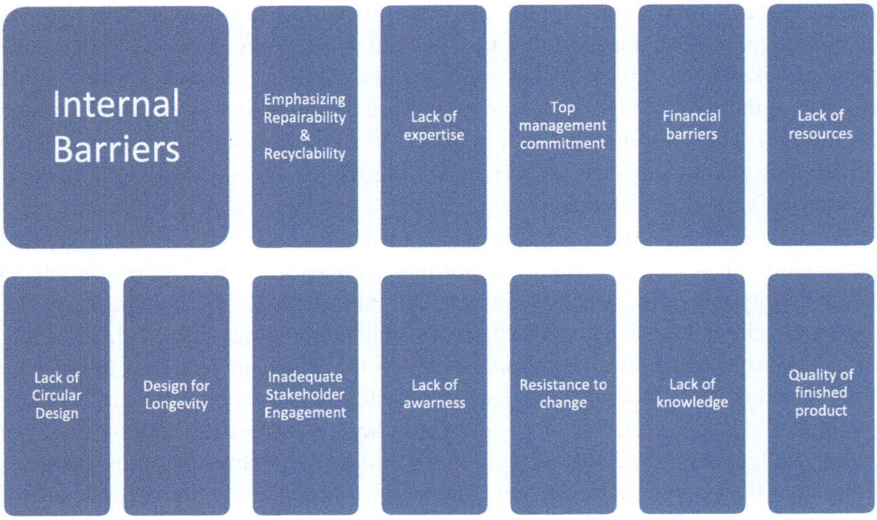

Fig. 4.1 Internal barriers

communication across different departments within the organization can impede the integration of circular practices into the entire supply chain. The following figure (Fig. 4.1) shows the various internal barriers in detail.

4.2.1 Lack of Circular Design

The absence of circular design principles is a notable internal barrier that organizations encounter when transitioning to Circular Supply Chain (CSC) practices. Circular design fundamentally shifts the traditional approach to product development by emphasizing the importance of longevity, repairability, and recyclability in the design process. This lack of focus on circular design can manifest in various ways, often leading to products that are not optimized for a circular economy (Lehtonen 2024). When products do not account for their entire lifecycle, organizations can face significant challenges at the end-of-life phase, where the disposal or recycling of these items becomes problematic.

4.2.2 Design for Longevity

Creating products designed for longevity is crucial in minimizing waste and maximizing resource utility. When products are built to last and withstand wear and tear over time, they not only reduce the frequency of replacement but also minimize environmental impacts associated with manufacturing and disposal (Kazancoglu et al. 2022). Products that are not designed with durability in mind often lead to higher consumption of materials and increased waste, exacerbating environmental issues.

4.2.3 *Emphasizing Repairability and Recyclability*

Furthermore, repairability is a vital component of circular design that allows consumers to extend the lifespan of a product rather than discarding it at the first sign of failure. By facilitating repair, organizations can reduce waste and retain value within the supply chain, aligning closely with the principles of a circular economy. In addition, emphasizing recyclability ensures that a product can be efficiently processed at the end of its life, allowing materials to be reclaimed and reused in new products (Chen et al. 2024). However, when products are not designed for easy disassembly or recycling, valuable materials can end up in landfills rather than being reintegrated into the production cycle.

The failure to integrate these principles often leads to increased waste and resource consumption. For instance, if a product is difficult to repair or recycle, consumers may choose to replace it rather than invest in maintenance or reclamation. This not only results in higher disposal rates but also diminishes the potential for recovering valuable materials from discarded products. Consequently, organizations may find themselves stuck in a linear production mindset, wherein the environmental consequences of their products are not adequately addressed, perpetuating a cycle of unsustainable consumption.

4.2.4 *Inadequate Stakeholder Engagement*

Effective stakeholder engagement is a critical component for the successful implementation of Circular Supply Chain (CSC) practices, as it ensures that various perspectives are considered and integrated into the decision-making process. Stakeholders include a wide array of groups such as employees, management, shareholders, suppliers, customers, and local communities; their involvement is essential for fostering a collective commitment to circularity. Engaging stakeholders helps organizations align their sustainability goals with stakeholder interests, ultimately driving successful circular economy initiatives. However, many organizations face significant challenges in garnering support from internal stakeholders, which can impede the transition towards circularity (Govindan and Hasanagic 2018). Internal stakeholders, including employees and management, may have different priorities and levels of understanding regarding the benefits of circular practices. This can lead to fragmented efforts and creates barriers that must be overcome to achieve shared goals. Moreover, a lack of awareness or education about the principles of circularity may contribute to resistance and apathy, further complicating the engagement process.

A fundamental issue that many organizations experience is inadequate communication and collaboration among stakeholders. Insufficient transparency can create misunderstandings and misaligned priorities, leading to an environment where stakeholders are hesitant to participate fully in the transition efforts. This disengagement can result in a lack of commitment and investment in circular practices, as stakeholders may not see how these initiatives align with their interests or objectives (Khandelwal and Barua 2024). Effective communication is essential to clarify roles, expectations, and the shared benefits derived from implementing circular strategies. To navigate these challenges successfully, organizations must actively involve relevant stakeholders in the transition towards CSC practices. This requires establishing platforms for dialogue, fostering open communication, and facilitating collaboration among various stakeholder groups. By emphasizing the joint benefits of adopting circular practices—such as cost savings, resource efficiency, and enhanced brand reputation—organizations can effectively motivate stakeholders to take part in the initiative.

4.2.5 Lack of Expertise

A major perceived bottleneck of CE is the lack of relevant expertise or detail technical knowledge in place for the transition from linear to CE. Adopting circular practices involves a shift from a traditional linear economy to a circular one. Organizations need expertise in the principles of a circular economy, such as designing products for durability, reparability, and recyclability (Mangla et al. 2018). Without a deep understanding of these concepts, companies may struggle to implement sustainable practices. Implementing circular supply chain solutions often requires the use of advanced technologies, such as smart tracking systems, data analytics, and innovative recycling processes. A lack of expertise in these technologies can hinder the efficient execution of circular strategies. Circular supply chains often require collaboration with various stakeholders, including suppliers, customers, and recycling facilities. Lack of expertise in building and managing such collaborative networks can impede the successful implementation of circular practices.

4.2.6 Lack of Awareness

The "lack of awareness" is a significant barrier that organizations may encounter when attempting to implement circular supply chain practices. This barrier refers to a limited understanding or recognition within an organization, its stakeholders, or the broader industry about the principles, benefits, and importance of adopting circular economy practices in the supply chain. Many stakeholders may not be familiar with the fundamental principles of a circular economy, which involves minimizing waste, promoting resource efficiency, and designing products for reusability and recyclability (Sadid et al. 2023). Without a clear understanding of these concepts, decision-makers may not recognize the value of transitioning to a circular supply chain. Lack of awareness regarding the environmental impacts of traditional linear supply chains may lead organizations to underestimate the importance of adopting more sustainable practices. Understanding the ecological consequences of waste generation and inefficient resource use is crucial for motivating change. Stakeholders may be unaware of the potential economic, social, and environmental benefits associated with circular supply chain practices. These benefits include cost savings, reduced environmental footprint, enhanced brand reputation, and new business opportunities. Without this awareness, organizations may be hesitant to invest in circular initiatives. If customers are not informed about the environmental impact of their choices and the benefits of circular products, there may be limited demand for such offerings, affecting the market viability of circular supply chain practices (Singh and Maheswaran 2024).

4.2.7 Top Management Commitment

Many researchers emphasize the role and vision of top managers' appreciation of new ideas and changes that can transform the way enterprises operate or design their supply chains. If there is any resistance from managers to change their corporate strategy, successful implementation of CE initiative cannot be realized. To overcome this barrier, top managers can conduct workshops and training of their workers and suppliers to promote CE practices in their enterprise and supply chain (Farooque et al. 2019). This barrier refers to a situation where senior executives and top-level management within an organization do not fully endorse or prioritize the adoption of circular supply chain practices. Without strong commitment from top management, it may be challenging to allocate resources, secure necessary investments, and drive the cultural and operational changes required for a successful transition to a circular model. Lack of awareness, competing priorities, or a perception that circular practices may not align with short-term financial goals are common reasons for a lack of top management commitment.

4.2.8 Resistance to Change

Resistance to change is the reluctance or opposition from individuals within the organization to adopt new ways of working, particularly those associated with circular supply chain practices. It can lead to delays, implementation hurdles, and a failure to fully integrate circular principles into the supply chain. It can be rooted in a fear of the unknown, concerns about job security, or a preference for established routines (Kumar et al. 2024). The introduction of circular practices may disrupt existing workflows, require new skills, or challenge established norms. Employees may resist change due to a lack of understanding, fear of job displacement, or a perception that circular initiatives are impractical.

4.2.9 Financial Barriers

Economic factors significantly hinder organizations from adopting Circular Supply Chain (CSC) practices, acting as substantial external barriers. These barriers can manifest in various ways, including financial constraints, volatile market conditions, and the overall economic environment (Lehtonen 2024). These economic challenges can complicate the transition towards more sustainable models, compelling organizations to prioritize short-term profits over long-term sustainability. One prominent economic barrier is budgetary constraints faced by many organizations, especially small and medium-sized enterprises (SMEs) (Khandelwal and Barua 2024). Implementing circular practices often requires initial investments in new technologies, infrastructure, and processes, which can be prohibitively expensive. As many SMEs operate with limited financial resources, the high costs associated with transitioning to a circular economy can deter them from adopting necessary circular practices. Without adequate financial backing, these organizations may struggle to justify the costs involved in implementing sustainable initiatives. Fluctuating material prices also present significant challenges for organizations aiming to implement circular practices. Price volatility can be particularly detrimental in industries heavily reliant on raw materials, where unpredictable market prices increase the financial risks associated with sourcing and using recycled or reused materials.

4.2.10 Lack of Knowledge and Technical & Technological Capacity

The lack of technical and technological capacity of enterprises as well as their inability to have possible means to embrace CE practices. CE is technologically centered and technology is a requisite in CE implementation. For instance, available technology that can provide real-time data can enhance the opportunity for a planned predictive maintenance scheme. Currently, despite the health hazards as well as environmental risks, many enterprises use low-grade technologies and unskilled workers to manage collections of returned parts in the industry. The internal barrier of lack of knowledge and technology was mentioned most by firms with the circular supplies model or the resource recovery model (Ayati et al. 2022). For firms with one of these models, product development from renewable, recyclable, biodegradable, or direct waste was key to their business. For the resource recovery model, this shortage of knowledge and technology specifically concerned recycling for the recovery of waste. Sometimes, machinery had to be adjusted or new knowledge or technology had to be developed. Related to the circular supplies model, the procurement of circular materials also required a change in product development processes; it was perceived as difficult since knowledge and/or technology were lacking (Mehmood et al. 2021).

4.2.11 Lack of Resources (Singh and Maheswaran 2024)

Lack of various resource capabilities hinders automobile enterprises from recycling, reuse and recycling. Unlike the conventional linear material product flow, the initial stage of developing a closed loop may require investment to be made by enterprises to achieve circular products. Such resources including organizational resource, financial resources, access to capital and availability of public funds for CE implementation may be unavailable, scarce and inaccessible.

4.2.12 Quality of Finished Product

Although product quality is identified as a driver of CE implementation, it is also perceived that product quality as a significant barrier to CE implementation and consider the negative perception of circular products as poor quality to limit its market demand (Souza Piao et al. 2024).

4.3 External Barriers

External barriers, on the other hand, are challenges that come from outside the organization and are often beyond its immediate control. Regulatory frameworks and government policies can either facilitate or hinder the transition to a circular supply chain. Inconsistent or unclear regulations related to recycling, waste management, and environmental standards may pose significant hurdles (Taddei et al. 2024). External factors such as market demands, customer preferences, and economic conditions can influence the feasibility and acceptance of circular products as shown in Fig. 4.2. Limited availability of recycling infrastructure and services in the external environment may also be a barrier, as efficient reverse logistics and waste management systems are crucial for closing the loop.

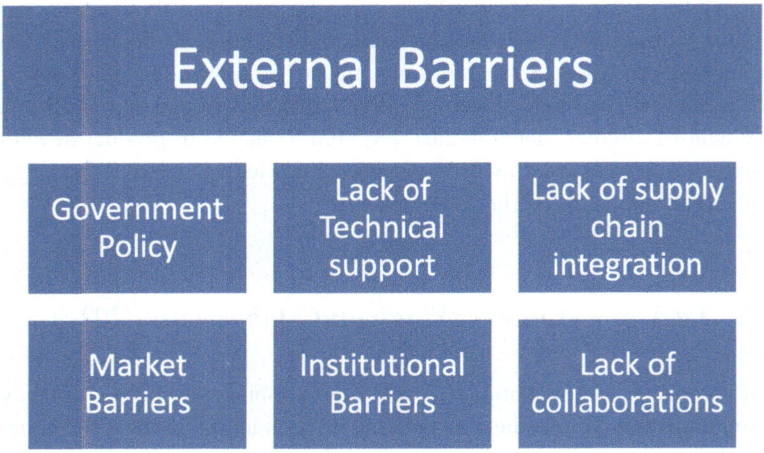

Fig. 4.2 External barriers

4.3.1 Government Policies

Government policies play a crucial role in shaping the business environment and influencing the decisions of organizations. When there is an absence of or inadequate policies that support circular economy principles, companies may face challenges in implementing sustainable and circular practices (Upadhyay and Shukla 2024). Also, government agencies that support the industry to make the transition are less resourceful with inadequate technical capacity to carry out their duties effectively as well as lack effective. The absence of clear and comprehensive regulations pertaining to circular economy practices can create uncertainty for businesses. Regulations related to waste management, product design, recycling standards, and environmental sustainability are essential to guide companies in adopting circular supply chain initiatives. Without a regulatory framework that encourages or mandates circular practices, companies may lack the necessary guidance and motivation to invest in sustainable and circular approaches. On the other hand, Government incentives, subsidies, and financial support can significantly influence the decision-making of businesses. Lack of financial support or tax incentives for adopting circular practices may make it economically challenging for companies to invest in sustainable initiatives (Sadid et al. 2023). Finally, the establishment of standardized practices and certification systems is essential for ensuring consistency and credibility in circular supply chain efforts. The absence of clear standards may hinder the development and adoption of circular economy practices. Lack of standardized guidelines and certifications can result in a fragmented landscape, making it challenging for businesses to communicate their circular achievements and for consumers to make informed choices.

4.3.2 Lack of Technical Support

These barriers organisation may encounter when attempting to implement circular supply chain practices. This barrier refers to a deficiency in the availability of technical expertise, knowledge, and support systems that are essential for the successful design, implementation, and optimization of circular processes throughout the supply chain. Transitioning to a circular supply chain often involves complex processes, such as recycling, remanufacturing, and reverse logistics. Implementing these processes requires specialized technical knowledge and expertise. The complexity of circular practices can be daunting for organizations that lack the technical know-how (Singh and Maheswaran 2024). Without adequate technical support, companies may struggle to design and implement effective circular processes. Circular supply chains often require the adoption of innovative technologies, such as advanced recycling processes, sustainable materials, and eco-design tools. Lack of technical support in adopting and integrating these technologies can be a barrier. Organizations may be hesitant to invest in innovative technologies without the guidance and technical support needed for successful implementation.

4.3.3 Lack of Supply Chain Integration and Effects of Supply Chain Complexity

Modern enterprises operate in the context of complex supply chains which involve the need to take a decision on a range of factors. Since CE implementation may even require the need to close the loop of the traditional supply chains, dynamic complexity and deep uncertainty is even much more significant in such a scenario. Supply chain dependencies were found to be problematic in three CBMs: circular supplies, resource recovery, and product life extension. For the circular supplies model, the problem was mainly caused by the limited number of suppliers of circular materials (Souza Piao et al. 2024). Since the circular economy is still in its infancy, few suppliers are already producing biodegradable or recyclable materials. These barriers can significantly impact the efficiency, transparency, and sustainability of supply chain operations. Supply chain integration involves the seamless coordination and collaboration of various components within the supply chain, including suppliers, manufacturers, distributors, and retailers. In the context of circular supply chains, integration is essential for the effective implementation of closed-loop systems, reverse logistics, and circular product life cycles. When there is a lack of integration, information flow and coordination between different stages of the supply chain may be disrupted.

This can result in inefficiencies, increased lead times, and difficulties in implementing circular practices, such as product take-back, remanufacturing, and recycling. Supply chain complexity refers to the intricacies and challenges associated with managing a network of suppliers, manufacturers, distributors, and other stakeholders. In a globalized and dynamic business environment, supply chains often become complex due to factors such as diverse product lines, geographic dispersion, and evolving market demands (Luthra et al. 2022). The complexities of a supply chain can complicate the implementation of circular practices. For instance, coordinating the return of used products for recycling or remanufacturing becomes more challenging in complex supply chains. The lack of visibility and control over the entire chain can hinder efforts to close the loop effectively. Circular supply chains often require collaboration among various stakeholders, including suppliers, manufacturers, logistics partners, and recyclers. A lack of collaboration can impede the sharing of information and resources necessary for the successful implementation of circular practices. Without effective collaboration, it becomes difficult to establish closed-loop systems, manage reverse logistics efficiently, and coordinate efforts to minimize waste and resource consumption throughout the supply chain (Lehtonen 2024).

4.3.4 Market Barriers

Market dynamics and consumer behaviour significantly influence the adoption of Circular Supply Chain (CSC) practices. These external barriers can create substantial challenges for businesses seeking to implement circular strategies while pursuing sustainability goals. As awareness of sustainability grows among consumers, there is mounting pressure on businesses to align with these expectations and innovate towards more sustainable practices (Chen et al. 2024). However, deeply entrenched consumer preferences can hinder this transition, particularly when traditional consumption patterns Favor convenience and immediate gratification over sustainable choices. The increasing awareness of sustainability among consumers drives a demand for more environmentally friendly products and practices. As a result, businesses are compelled to adopt circular practices to remain competitive in the marketplace. However, this shift often requires significant changes in operations, supply chains, and product designs, which can be resource-intensive and complex. Given the urgency for businesses to respond to consumer demands, they must balance these pressures with potential challenges that come with redefining their operational frameworks.

Despite the growing pressure to adopt circular practices, existing consumer preferences can complicate this transition. Many consumers exhibit a strong inclination towards convenience and the familiarity of traditional ownership models rather than embracing new, sustainable alternatives. This attachment to conventional consumption patterns can create resistance to adopting more circular models, especially when such models require changes in behaviour, such as sharing products instead of owning them outright. A further challenge stems from consumer perceptions regarding the value and quality of circular products in comparison to their traditional counterparts. If consumers do not view circular products as delivering equal or superior quality, they may be reluctant to choose them over conventional options.

This perception can arise from a lack of understanding of the benefits of circularity, or from previous experiences with products that did not meet quality expectations. Companies, therefore, face a dual challenge of both promoting the attributes of their circular offerings and educating consumers on the advantages of these products in terms of sustainability and performance.

4.3.5 *Institutional Barriers*

For the resource recovery model, institutional barriers were related to legislation on the use of waste as a resource. Multiple firms mentioned they wanted to experiment with waste to see if it could be used as a resource but were hampered by legislation. Another perceived problem in addition to entrenched rules were not focused on a circular economy (Lehtonen 2024). For example, KPIs and accounting rules were still very focused on a linear economy. The circular supplies and product life extension models seemed less influenced by these barriers. Some barriers mentioned by the firms were not directly related to a specific CBMs but instead were related to the infancy of the circular economy. Currently, a small number of actors are active in the circular economy, which results in low legitimacy for circular products and CBMs. A lack of awareness in society was one of the barriers mentioned. Another institutional barrier was the difficulty of finding investors willing to invest in firms with a product-as-a-service model. Banks were not perceived as willing investors by some of the firms.

4.3.6 *Lack of Collaboration and Partnerships*

Collaboration and partnerships are fundamental in establishing effective Circular Supply Chain (CSC) practices, as they enable organizations to leverage shared resources and knowledge (Kazancoglu et al. 2022). By working together with suppliers, customers, and other stakeholders, businesses can optimize processes, reduce waste, and enhance overall sustainability. These collaborative efforts can drive innovation and lead to the development of solutions that align with the principles of a circular economy, ultimately benefiting all parties involved. Despite the critical importance of collaboration, external barriers frequently hinder these efforts in implementing CSC practices. Organizations often encounter challenges such as financial constraints, regulatory hurdles, and cultural resistance, which can impede the formation of effective partnerships. These external factors may discourage organizations from engaging in collaborative initiatives, leading to missed opportunities for resource sharing and waste reduction. Establishing mutual trust and cooperation among stakeholders is essential for successful collaboration.

However, in competitive markets, organizations may struggle to foster such trust, particularly when sharing information or resources is perceived as risky. Companies might fear that disclosing sensitive data could benefit competitors or lead to unfavourable consequences for their own operations (Amiri et al. 2022). As a result, the potential for collaboration may be stifled by fears of competitive disadvantage, undermining the collective success of circular initiatives. Information sharing is a critical component of successful collaboration in a circular supply chain. Organizations need to exchange insights and data to effectively coordinate activities and optimize resource utilization. However, concerns about confidentiality and the competitive implications of sharing critical operational data often create barriers to open communication. As a result, organizations may miss vital opportunities to enhance their collaborative efforts, ultimately hindering the success of their CSC initiatives.

4.4 Strategies to Overcome These Barriers

Overcoming barriers to the implementation of circular supply chain practices requires a comprehensive and strategic approach. Here are strategies to address some of the common barriers mentioned:

Conduct education and awareness programs at all levels of the organization. Provide training sessions to employees, management, and other stakeholders to increase understanding of circular economy principles and the benefits of adopting circular supply chain practices. Invest in training and development programs to enhance the technical skills and knowledge of employees (Luthra et al. 2022). Collaborate with external experts and consultants to bring in specialized expertise. Foster a culture of continuous learning and innovation within the organization. Engage top management in discussions about the long-term benefits of circular supply chains. Clearly articulate the business case, including potential cost savings, improved brand reputation, and environmental benefits. Implement change management strategies to address resistance, including communication plans, employee involvement, and gradual implementation. Conduct a thorough cost-benefit analysis to demonstrate the long-term financial advantages of circular practices.

Explore financing options, such as government grants, subsidies, or partnerships. Emphasize the potential return on investment, efficiency gains, and market competitiveness associated with circular supply chain initiatives. Advocate for supportive policies and regulations by actively engaging with policymakers. Collaborate with industry associations and other organizations to present a unified voice in support of circular economy initiatives. Participate in public-private partnerships to influence the development of regulatory frameworks that encourage sustainable practices (Ada et al. 2021; Kazancoglu et al. 2022). Collaborate with research institutions, industry associations, and external experts to access technical support. Invest in research and development activities to drive innovation in circular supply chain technologies. Foster partnerships with organizations that specialize in providing technical assistance and expertise. Invest in advanced technologies, such as supply chain management systems and blockchain, to enhance visibility and integration across the supply chain. Streamline supply chain processes to reduce complexity.

Foster collaboration through clear communication channels, partnerships, and information sharing. Establish strong relationships with key stakeholders, including suppliers, manufacturers, and recycling partners. Clearly communicate the benefits of collaboration and shared goals. Implement collaborative platforms and technologies to facilitate communication and information sharing throughout the supply chain. Stay informed about market trends, consumer preferences, and regulatory changes. Engage with consumers to understand their expectations and educate them about the benefits of circular products (Upadhyay and Shukla 2024). Conduct pilot projects to test market acceptance before full-scale implementation. Advocate for government incentives and subsidies by actively participating in policy discussions. Clearly articulate the positive economic and environmental impacts of circular practices. Collaborate with industry associations to lobby for supportive financial measures.

To overcome circular design integration barriers, organizations must enhance their design processes by embedding circularity from the outset. This necessitates comprehensive training for design teams on circular economy principles and methodologies, ensuring that they are equipped to create products suited for the demands of a sustainable future. Shifts in design methodologies may include incorporating life cycle assessments during the design phase, collaborating with stakeholders for feedback on product usability and recyclability, and adopting best practices for material selection and sourcing. By prioritizing circular design, organizations can more effectively contribute to a sustainable economy and create products that are more resilient and environmentally friendly (Chen et al. 2024; Singh and Maheswaran 2024).

By engaging stakeholders meaningfully, organizations can build a shared vision and collective ownership of circular initiatives. This collaborative approach not only aligns the different interests of stakeholders but also encourages a sense of accountability and responsibility, which is crucial in achieving sustainability objectives. When stakeholders can see the mutual advantages of transitioning to circular practices, their commitment strengthens, leading to more robust support and a higher likelihood of successful implementation of CSC practices. Overcoming the challenges of stakeholder engagement is essential for organizations aiming to implement CSC practices effectively, and fostering an inclusive, communicative environment is vital in this process. Market dynamics and consumer behaviour present significant external barriers that complicate the implementation of CSC practices (Govindan and Hasanagic 2018; Luthra et al. 2022).

Businesses must navigate these challenges by effectively communicating the value of circular products, addressing consumer preferences, and fostering a cultural shift towards sustainability. Engaging consumers through education, transparency, and demonstrating the tangible benefits of circularity can help overcome these barriers, facilitating a smoother transition to circular supply chains. The challenges presented by competitive market dynamics, organizations must actively work to cultivate a culture of collaboration and trust among stakeholders. This can involve implementing transparent communication strategies, fostering a shared vision, and emphasizing the long-term benefits of joint efforts. By demonstrating the value of collaboration and actively addressing concerns related to information sharing, businesses can build stronger partnerships that facilitate resource sharing and coordination in their circular supply chains (Mehmood et al. 2021). While collaboration and partnerships are crucial for establishing effective CSC practices, external barriers, concerns about trust, and competition must be navigated carefully to unlock their full potential. Organizations that successfully foster collaborative relationships will be better positioned to drive sustainable outcomes in their supply chains.

4.5 Conclusion

The adoption of Circular Supply Chain (CSC) practices presents a transformative opportunity for businesses to achieve environmental sustainability and long-term economic viability. However, the journey towards implementing CSC models is fraught with numerous internal and external barriers that organizations must overcome to realize their sustainability goals. The transition from traditional linear supply chains, which follow the 'take-make-dispose' model, to circular models requires a comprehensive restructuring of operations, technologies, and strategies. In this context, the identification and management of these barriers become essential for the successful implementation of CSC. One of the most significant internal barriers to CSC adoption is the lack of expertise within organizations. The principles of circularity, such as designing for durability, repairability, and recyclability, require specialized knowledge that many companies may not possess. This lack of technical and technological capacity often hinders organizations from fully integrating circular processes into their supply chains. Moreover, the absence of awareness and education about circular practices among employees and management further exacerbates this issue. When decision-makers do not understand the economic and environmental benefits of circularity, there is a reluctance to invest in necessary changes. Investing in new technologies, redesigning products for circularity, and establishing recycling or remanufacturing facilities require substantial capital. Moreover, the financial business case for CSC may not always be clear, making it difficult for organizations to justify these investments to stakeholders and shareholders.

In addition to internal challenges, external barriers, including governmental regulations, market dynamics, and supply chain complexities, significantly impact the adoption of CSC practices. Regulatory frameworks often lag behind the needs of circular economy initiatives. Inconsistent or unclear policies related to recycling, waste management, and product design can create uncertainty for businesses trying to implement circular practices. Moreover, the lack of government incentives, such as subsidies or tax breaks for sustainable initiatives, discourages companies from adopting circular models. Market dynamics, particularly consumer behaviour, also present a significant challenge. While awareness of sustainability is increasing, many consumers are still hesitant to embrace circular products due to concerns about quality or perceived inconvenience (Kumar et al. 2024). The shift from traditional ownership models to circular alternatives, such as product-as-a-service, requires a change in consumer mindset. Companies must educate their customers about the value and benefits of circular products to foster greater acceptance and demand. Supply chain complexity and the lack of integration among different stakeholders further complicate CSC implementation. Circular supply chains require close collaboration between suppliers, manufacturers, retailers, and recyclers to ensure that materials and products are continuously reused and recycled. However, the global nature of modern supply chains often creates barriers to coordination and transparency. For example, companies may struggle to track products throughout their lifecycle, making it difficult to implement effective take-back or recycling programs (Upadhyay and Shukla 2024).

Overcoming these barriers requires a multi-faceted approach. Internally, organizations must invest in education and training programs to enhance the technical expertise of their employees and management teams. By fostering a culture of continuous learning and innovation, companies can equip their workforce with the skills necessary to implement and maintain circular practices. Top management plays a critical role in driving these changes, and leadership must be fully committed to the long-term benefits of CSC. Engaging top management through workshops, training sessions, and clear communication about the potential cost savings and environmental benefits of circular practices can help overcome resistance to change. Externally, businesses must advocate for supportive government policies and regulatory frameworks that encourage circular economy initiatives. Engaging with policymakers and participating in industry associations can help create a more favourable environment for CSC adoption (Shaikh et al. 2024). Additionally, companies should explore public-private partnerships and other collaborative efforts to access technical support and financial resources for circular initiatives. In conclusion, while the barriers to adopting Circular Supply Chain practices are significant, they are not insurmountable. Through a combination of internal and external strategies—ranging from education, leadership engagement, and financial investment to collaboration, regulatory advocacy, and consumer engagement—organizations can successfully transition to circular supply chains (Sadid et al. 2023). By doing so, businesses can not only meet growing consumer, and stakeholder demands for sustainability but also position themselves for long-term economic success in a rapidly evolving market.

References

Ada N et al (2021) Analyzing barriers of circular food supply chains and proposing industry 4.0 solutions. Sustain For 13(12):6812

Amiri M et al (2022) Evaluating barriers and challenges of circular supply chains using a decision-making model based on rough sets. Int J Environ Sci Technol 19:1–22

Ayati SM et al (2022) Toward a circular supply chain: understanding barriers from the perspective of recovery approaches. J Clean Prod 359:131775
Chen G, Geng D, Zhu Q (2024) Implementation barriers to the circular supply chain mode for pre-consumer textile waste reprocessing in China. Resour Conserv Recycl 206:107621
Farooque M, Zhang A, Liu Y (2019) Barriers to circular food supply chains in China. Supply Chain Manag Int J 24(5):677–696
Govindan K, Hasanagic M (2018) A systematic review on drivers, barriers, and practices towards circular economy: a supply chain perspective. Int J Prod Res 56(1–2):278–311
Kazancoglu I et al (2022) Investigating barriers to circular supply chain in the textile industry from stakeholders' perspective. Int J Log Res Appl 25(4–5):521–548
Khandelwal C, Barua MK (2024) Prioritizing circular supply chain management barriers using fuzzy AHP: case of the Indian plastic industry. Glob Bus Rev 25(1):232–251
Kumar A, Mangla SK, Kumar P (2024) Barriers for adoption of industry 4.0 in sustainable food supply chain: a circular economy perspective. Int J Product Perform Manag 73(2):385–411
Lehtonen S (2024) Circular supply chains in Finland: drivers, enablers, and barriers. LUT University
Luthra S et al (2022) Overcoming barriers to cross-sector collaboration in circular supply chain management: a multi-method approach. Transp Res Part E Logist Transp Rev 157:102582
Mangla SK et al (2018) Barriers to effective circular supply chain management in a developing country context. Prod Plan Control 29(6):551–569
Mehmood A et al (2021) Drivers and barriers towards circular economy in agri-food supply chain: a review. Bus Strategy Dev 4(4):465–481
Sadid MS et al (2023) Assessing barriers to circular supply chain management in the plastic industry of Bangladesh. In: Proceedings of the 6th industrial engineering and operations management Bangladesh conference Dhaka, Bangladesh, December 26–28, 2023
Shaikh AR, Qazi AA, Appolloni A (2024) Identification and evaluation of the contextual relationship among barriers to the circular supply chain in the Pakistani context–an interpretive structural modelling approach. Prod Plan Control 35(10):1148–1163
Singh PK, Maheswaran R (2024) Analysis of social barriers to sustainable innovation and digitisation in supply chain. Environ Dev Sustain 26(2):5223–5248
Souza Piao R et al (2024) Barriers toward circular economy transition: exploring different stakeholders' perspectives. Corp Soc Responsib Environ Manag 31(1):153–168
Taddei E et al (2024) Circular supply chains theoretical gaps and practical barriers: a model to support approaching firms in the era of industry 4.0. Comput Ind Eng 190:110049
Upadhyay A, Shukla AC (2024) Circular supply chain management: content analysis with specific drivers and barriers. Int J Ind Eng 31(1)

Chapter 5
The Smart Manufacturing Concept

Abstract The concept of smart manufacturing, as outlined in this document, revolves around the integration of innovative technologies and methodologies that enhance production efficiency and promote sustainable industrial practices. Key driving forces behind this transition include the rise of climate change regulations, escalating raw material costs, and the environmental consequences of traditional waste disposal methods. Smart manufacturing emphasizes the circular economy, which shifts production from a linear model ("cradle to grave") to a more sustainable circular model ("cradle to cradle"), where products and resources are continuously reused, remanufactured, or recycled. This approach not only extends the lifecycle of goods but also minimizes waste and maximizes resource utilization, contributing to long-term sustainability. By adopting business models that focus on reparability, durability, and recyclability, industries can significantly reduce their dependency on raw materials and contribute to environmental preservation. The integration of circular supply chains further enables businesses to maintain product value and efficiency throughout their lifecycle.

Keywords Circular economy · Smart manufacturing · Sustainable business models · Supply chain management · Resource efficiency · Eco-innovation

5.1 Introduction

Over the past decade, companies have systematically implemented business strategies aimed at prolonging the lifespan of goods, their components, and beneficial waste outputs. This response is driven by climate change regulations, escalating raw material acquisition costs, and the environmental impacts of by-products and waste disposal processes (Amjad and Diaz-Elsayed 2024). By reusing, remanufacturing, and recycling goods, components, trash, and by-products as long as feasible back into commercial cycles, business actions in this area often entail the conservation of materials. The circular economy, advocating for manufacturing methods that are inherently restorative and shifting product value chains from linear ("cradle to

J. Kandasamy et al., *Circular Supply Chain Management*, Circular Economy and Sustainability, https://doi.org/10.1007/978-3-031-81750-2_5

grave") to circular ("cradle to cradle") cycles, considers this type of market movement as a fundamental component (Atif 2023). Thus, the capacity of business models to create circular value chains that optimize resource efficiency is a key component of circular economy business models. This can be achieved by cutting back on primary extraction procedures and eliminating disposal operations that cause priceless resources to escape the economy. Long-term material utilization is necessary to reduce primary extraction and disposal activities. This may be accomplished by using business models that focus on designing items with improved features for reparability and durability. Alternative business models entail the use of restorative procedures that generate value via the recycling of waste materials and byproducts as well as through product remanufacturing, or renewing of items. In a broader sense, business models based on the circular economy may comprise intricate networks of companies that constantly trade resources to create new economic value. This is made possible by cutting-edge technology, supply chain ecosystems that support product-service offers, and industrial symbiosis programs that unite businesses from many economic sectors (Bressanelli et al. 2022b). The idea of the circular economy is becoming more and more well-known in policy, practitioner, and academic circles, but its practical implementation is still sparse and unstable. A more thorough knowledge of "circularity" qualities of enterprise models is necessary for shifting to a circular economy centered on renewing design, manufacturing incorporating inverse cycles, cascading processes, and cross-sector partnerships outside conventional supply chain boundaries. In this context, it is also important to have a deeper understanding of the role that "circular" supply chain activities play in enabling circular economy business models, given their growing complexity and scale (Bressanelli et al. 2022a). More precisely, the advantages of innovative business ventures in the circular economy are being more acknowledged, as is their capacity to spur productivity and growth on the backs of environmental, social, and economic sustainability imperatives.

5.2 The Eco-Innovative Elements That Embody the "Circularity"

The components of enterprise models within the circular economy, along with the key attributes and enabling functions of "circular" supply chains, are areas that have been largely unexplored (Amjad and Diaz-Elsayed 2024). In this context, several critical questions are emerging:

1. What are the prominent "circularity" characteristics of business models that implement circular economy principles?
2. How do these characteristics facilitate the continuous flow of resources?
3. What enabling functions do "circular" supply networks serve?
4. What essential qualities define an archetype of a circular supply chain?

Exploring these questions is vital for understanding and advancing the effectiveness of circular economy practices in business and supply chain management. A rising corpus of work is defining the theoretical and practical underpinnings of the circular economy's conceptual paradigm, which views sustainability as a fundamental component of both production processes and the economy at large (also known as the "triple bottom line"). The market logic that drives businesses and their economic operations is heavily reliant on the sustainability capabilities of organizations.

Nonetheless, an exceptional confluence of technological, political, and social elements is permitting a smooth and efficient shift towards a circular economy. Innovations in business models that optimize societal and environmental advantages without compromising commercial gains are being made possible by the current state of the economy. In the circular economy, closed-loop systems play a central role by minimizing waste through practices like reusing, repairing, remanufacturing, and recycling (Blunck and Werthmann 2017). This approach prioritizes delivering functionality and experiences over product ownership, promoting resource efficiency. Additionally, management strategies based on shared or collaborative consumption models are gaining traction, fostering sustainable resource utilization and reducing environmental impact.

The elements may be converted into useful characteristics of business model innovations, the primary goal of which is to increase the longevity of products (Bressanelli et al. 2022b). This can be accomplished in the following ways: (1) goods are sold as services, prioritizing "utilization value" over "exchange value". (2) goods are reused, repaired, or remanufactured; (3) shared goods utilization results in higher materials efficiency; and (4) stock value is maintained through service-life extension activities. These features essentially embody the restorative and regenerative qualities of business models, showcasing their capacity to regenerate materials to a typically improved condition and restore vitality to the environment. Since both ideas include "recuperation," or recovering resources for additional use, we will simplify the definition by using the term "restorative" to also refer to an organization's capacity for "regenerative" supply chain activities. By distinctness, a circular economy is a restorative economic system, meaning that resources, products, and components are retained in the economy for as long as possible at their greatest usability and worth (Chowdhury et al. 2023). The business aspects are grounded on a basic idea that emphasizes the crucial necessity of firms' mending capacities. It also suggests that a company model's regenerative qualities might be intentionally created. More precisely, we suggest that the industry, process, and product levels may all intentionally be structured to maximize the restorative potential of circular economy business models (Bressanelli et al. 2022a). Therefore, via connections to industry, processes, and restorative aspects of goods, we identify the following three tiers of company "circularity" capabilities:

1. At the product level: Reparability, durability, upgradability, and recyclability are examples of physical properties of items that provide life extension and restoration.

2. At the company level, restorative procedures such as recycling, remanufacturing, refurbishing, reconditioning, and reusing are outlined by the All-Party Parliamentary Sustainable Resource Group as follows:
   (a) Reusing: Utilizing a product again without any alterations.
   (b) Repairing: Fixing a product without providing a warranty for the overall product.
   (c) Reconditioning: Making modifications to a product's parts to restore its functionality, though not necessarily to a "like-new" state.
   (d) Refurbishing: Implementing significant cosmetic improvements to a product, potentially returning it to a "like-new" state with limited functional enhancements.
   (e) Remanufacturing: Undertaking a series of manufacturing operations on a product or part that has reached the end of its useful life, with the aim of restoring it to a "like-new" condition, which may include improved functionalities.
   (f) Recycling: Transforming a product's raw materials into new products.
3. In the commercial enterprise level: This level includes resource and infrastructure sharing, participation in waste and by-product synergy systems, recycling of used materials and renewable resources amongst businesses, and industrial symbiotic processes involving various organizations.

5.3 Supply Chain Business Model

The 1990s saw the rise of e-commerce along with new income streams, which led to the popularization of the business model idea. In this case, its original application was to quickly provide investors with concise yet thorough company concepts. To get at a consensus, several authors have provided varying definitions of the term, which have been thoroughly reviewed. Business models are simplified representations of the components within a complex organizational system and the relationships among them, employing comparative techniques to delineate their functions (Ciliberto et al. 2021). Amid increasing complexity, they define the organization's value proposition, value generation, and delivery mechanisms, as well as value capture strategies, aiming to facilitate analysis, planning, and effective communication. In most approaches, the organizational environment and value network are also taken into consideration, albeit to varying degrees.

The concept of circular business models (CBMs) has emerged from the integration of a practice-oriented approach to business model innovation and the imperatives of implementing the Circular Economy. CBMs are characterized by their integration of elements aimed at slowing, narrowing, and closing resource loops, thereby minimizing waste and emissions exiting the system while enhancing resource input into the organization and its value network (Cioffi et al. 2020). Furthermore, it's essential to underscore the interconnections between CBMs and circular supply chain management (CSCM), as we will elaborate in the following section. These connections may be made in many ways to closed loops, such as shutting, delaying, intensifying, narrowing, and dematerializing loops. Both dematerializing and amplifying loops.

The value production and distribution component, and specifically the supply chain, is where traditional enterprise models and those created for the circular economy diverge the most. Circular supply chain management (CSCM) involves configuring and coordinating the supply chain to achieve goals such as closing, narrowing, delaying, intensifying, and dematerializing resource loops. Even though CSCM is crucial for CBMs and, consequently, for putting the Circular Economy into practice, there is still a lot to learn about this field of study. Furthermore, it's critical to place CSCM in the context of other closely similar but distinct ideas, such as green supply chain management (GSCM) or sustainable supply chain management (SSCM), where closed loops are not a major problem (Ćwiklicki and Wojnarowska 2020). The current study intends to contribute to the emergent tract of circular supply chain management (CCM) by offering industry perspectives on CBM and by proposing a framework for integrating CSM and CBM into sustainable development.

The contemporary concept of the Circular Economy draws from various schools of thought such as Cradle to Cradle, the Laws of Ecology, the Looped and Performance Economy, Regenerative practices, Industrial Ecology, Biomimicry or the Blue Economy. Its aim is to minimize waste and preserve the intrinsic value of products by recycling materials and/or energy after the use phase, thereby reducing resource consumption and preventing leakage out of the system. To fully harness the potential of sustainable business models (SBMs) at the organizational level, three key components are necessary: sustainable value creation, proactive management of a broader range of stakeholders, and a long-term outlook. Circular economy business models are considered a subset of SBMs or a broader approach to them. Strategies such as closing, narrowing, delaying, intensifying, and dematerializing loops help minimize resource input and waste and emissions leakage, thus enhancing sustainability performance (Gómez et al. 2018). Intensive usage phases and the adoption of service and software solutions are integral to this approach, complementing initiatives such as recycling, efficiency enhancements, and use phase extensions as shown in Fig. 5.1.

5.3.1 The Strategy

A business plan is created using a business model as a framework. A strategy is a way to outperform rivals and obtain an edge over them. Value generation, delivery, capture, and proposition all need to work together to produce a cohesive system, and strategy is what makes this coherence logical. A plan must be put into practice. To do this, it must be ascertained which crucial procedures must be followed in order to produce and deliver value as well as which performance goals must be met in order to carry out the plan. Multiple layers of interconnected hierarchy should be

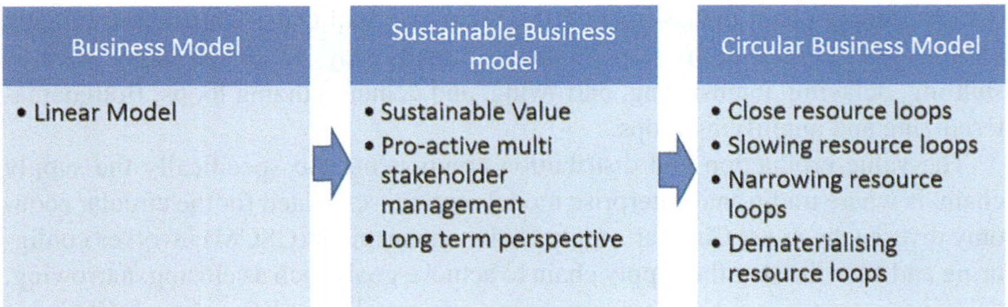

Fig. 5.1 Business model evolution

included in the core activities and performance targets (Ghaithan et al. 2023). These levels allow the organization's strategic plan to be translated into tactical and operational decision-making. Additionally, the levels give all levels the ability to report on, track, and manage their advancement toward the performance goals and their role in achieving strategic success.

5.3.2 *Procedure*

In a circular business model, the supply chain consists of eight key procedures: Plan, Source, Make, Deliver, Use, Return, Recover, and Enable. While these processes are also present in a linear business model, they require a shift in emphasis to effectively support a circular business model.

5.3.3 *Arrange*

The Plan procedure aims to assess the supply chain's needs and align them with available assets and resources. Traditionally, boundaries regarding natural resource usage were often overlooked as long as resources were accessible to meet supply chain demands. However, in a circular business model, supply networks must consider environmental constraints such as resource capacity and output exploitation. Therefore, a supply chain operating under a circular business model emphasizes the availability of natural resources and the impact of supply chain demands on these resources. Rooted in industrial ecology and natural ecosystems, the circular economy prioritizes environmental health. Consequently, integrating the availability of natural resources into the planning process of a circular business model is highly significant.

5.3.4 Source

The objectives of Source are to arrange for product delivery, accept and inspect the goods, transfer the product, and approve supplier payment. In a circular business model, the supply chain focuses on identifying and sourcing alternative materials as part of its sustainability initiatives. It is becoming more crucial to choose materials that reduce waste, are readily dismantled, recovered, and have a little impact on the environment.

5.3.5 Produce

The procedure Set goals for the manufacturing schedule, provide materials, manufacture and test, stage the product, and release the finished goods for delivery. The packaging and waste disposal sub processes that make up the Make process are anticipated to become more significant in a circular business model. When it comes to packing, the emphasis will be on low-cost, environmentally friendly materials that safeguard the goods not just during delivery but also during returns. In a circular business model, a notable aspect of the supply chain is waste management. Waste is generated during various stages, including extraction of natural resources for raw materials, manufacturing processes where unused components remain, and post-consumption when products are no longer needed. Effective waste management is vital as it closes the environmental cycle and supports sustainability efforts.

5.3.6 Provide

The subprocesses of the Deliver process encompass a range of activities, including but not limited to constructing loads, routing shipments, selecting carriers, receiving goods, picking, packing, documenting, shipping, and installing products. Deliver retail products, made-to-order products, stocked products, and engineer-to-order products. Deliver is seen as a crucial supply chain link in a circular business strategy. In addition to concentrating on product delivery to final consumers, Deliver will plan the distribution of replacement components for upkeep while the product is being used, so extending its lifespan. As a result, a subprocess must be added to the Deliver processes. Provide products for MRO (Maintenance, Repair, Overhaul).

5.3.7 Make Use Of

In a circular business model, consumption is a component of the supply chain, as defined by the terms "circular economy" and "circular supply chains". The end-user's usage or consumption of the product, along with any maintenance or repairs made to extend its lifespan, make up the use process. In both business-to-business and business-to-consumer markets, the end-user may be a single individual or an organization. Because these items are inputs for the process of making after recovery, a circular business model demands end users to return their products at the end of their useful lives. It is reasonable to anticipate that supply chains will incorporate the Use process given the existing ambiguity around the quantity, quality, and location of these returns. Better control over the volume, Caliber, and placement of return input is made possible by include the use process. Additionally, end users may be encouraged to maintain the product's quality and motivated to utilize it for as long as feasible.

5.3.8 Return

The goal of return is to reroute materials from the final consumer through the supply chain. Returns are considered one of the main processes since they play a significant role in supply chains. Moreover, take back following end-of-life is necessary for a circular business model, and returns are seen as a crucial element of any circular business model. Return is indeed a pivotal step in the supply chain of a circular business model. Product returns may occur due to defects, excess inventory, or the need for Maintenance, Repair, and Overhaul (MRO). Additionally, products that have reached the end of their useful life can also be returned in a circular business model. Therefore, the supply chain processes should include mechanisms for returning end-of-use items.

5.3.9 Recover

Recover is a process that should be included in a circular enterprise model's supply chain due to the significance of recovery operations. This study distinguishes between the following potential recovery processes: recycling, downcycling, upcycling, remanufacturing, and refurbishing. Table 4 provides a description of the process definitions. Additionally, some writers use terms like cannibalization, upgrading, and reconditioning. Upgrading is a choice wherein obsolete modules or parts are swapped out for ones that are more advanced technologically. This article does not use the term "upgrading" since it is defined similarly to "remanufacturing". Reconditioning is the process of surface, painting, and sleeving parts to return them to a suitable and/or functioning state. This is covered by the reuse definition. The selective removal of pieces from finished goods is known as cannibalization. This phrase is not used because the concepts of remanufacture, and recycling include the condition of cannibalization.

5.3.10 Enable

In a circular business model, the "Enable" procedure pertains to supply chain management. Unlike a linear business model, a circular model introduces distinct challenges for supply chain management. Material sourcing now considers both suppliers and end users, who dispose of items at the end of their useful lives. These end-of-life products can potentially serve as valuable resource sources when returned and recovered. It will be necessary to handle buy back and take back in order to facilitate the return of end-of-life items. While the number and quality of returns are unclear, it is crucial to minimize environmental effect, work efficiently and trace materials through numerous recovery channels. In the wake of these advancements, supply chain management within a circular business model will grow increasingly intricate and challenging to efficiently arrange.

5.4 The Performance Goal

Decoupling resource depletion from economic development is a circular economy's overarching performance goal. Three performance goals may be simultaneously achieved through a supply chain, which is known as the Triple Bottom Line, or 3P (People, Planet, Profit) (Gómez et al. 2018). Any supply chain can aim to achieve these performance goals. This indicates that a supply chain operating within a circular business model distinguishes itself by striving to achieve specific performance objectives. A circular business model's supply chain is defined by the following performance goals:

5.4.1 Reduce the Amount of Materials, Water, and Energy Used

Every step of the supply chain is subject to the performance goal of minimizing the usage of materials, water, and energy, which includes the procedure Get well. Recovering a product will need utilizing the least amount of resources, water, and energy feasible in order to make it available for further usage.

5.4.2 *Reduce the Amount of Inventory*

Without end-user demand, inventory shouldn't be on hand in a circular business model's supply chain. Production and supply need to occur exclusively in response to demand. Although the future consumption of raw materials, work-in-process, and finished goods is still undetermined, they have already used resources while they are accessible without demand. When resources are not converted to meet the need for consumption, items may become outdated due to shifts in demand. This performance goal suggests that Make-to-Order and Engineer-to-Order are better than Make-to-Stock in a supply chain inside a circular business model.

5.4.3 *Make the Most Effective Use of the Supply Chain's Assets*

This includes vehicles, warehouses, machinery, and equipment. Benefits to the economy and environment can be supported by using supply chain assets (including vehicles, warehouses, and machinery) more efficiently. The significance of raising the product's production or usage efficiency. It appears that this performance goal has something to do with the Toyota Production System's development of the Lean methodology. Eliminating non-value-adding tasks, improving quality, and providing value to end users are the goals of the lean methodology. It is acknowledged that the Lean methodology helps the achievement of environmental performance goals by encouraging the effective use of resources, including water, energy, and materials. Furthermore, the usage of sharing platforms to improve the utilization of available assets within supply chains is receiving greater attention in circular business models.

5.4.4 *Reduce the Amount of Waste*

Resources (materials, water, energy, and land) that are disposed of during or after resource extraction, production, and usage but are not beneficial to the final user are referred to as waste. Numerous writers emphasize how cutting waste may result in cost savings and how crucial it is to do so across the whole product life cycle. Throughout the whole life cycle of a product, waste arises at every stage: goods are made but not sold; they are sold but not returned; they are returned but not recovered; and, at the end, goods are recovered but neither employed as raw materials for new products nor resold to consumers.

5.4.5 *Maximize the Availability of the Product*

In a circular business model, the extended availability of a product for consumption reduces the resources required for its development over time, emphasizing longer product life cycles. Standardization and interoperability enable easy upgrades, modifications, maintenance, and repairs, further supporting longer product lifetimes. Additionally, encouraging more intensive product usage, such as through product sharing, ensures fewer items are needed overall, reducing the demand for fresh resources in product development and distribution.

5.4.6 *Increase as many Recovery Flows as Possible*

The objective of this performance goal is to enhance the frequency of product recovery across various supply networks or within the same supply chain. A product may lose some of its qualities after being recovered several times. The product could still be ideal for use in a different supply chain, though. This suggests that there would be no need to extract new materials, as the resources of the product would remain available for reuse.

5.5 A New Business Model Is Required for the Circular Supply Chain, Which Requires Collaboration

We examine related studies on collaborative innovation to acquire ideas on how it may be organized, as COI is a young field of study with little emphasis on collaborative innovation (Han et al. 2023). First, we offer variables that we extracted from the literature on open innovation and strategic management. Next, we illustrate the many forms of invention that fall under the umbrella of collaborative innovation. All along, references to COI are made.

5.5.1 *Cooperative Strategy Management*

In order to get a competitive edge, strategic management entails creating strategic visions, establishing goals, and creating, deciding upon, and putting into practice targeted tactics. Pursuing collaborative advantage is one tactic that may be used. It involves making conscious decisions and choosing certain techniques, but it can be a valuable strategic tool and a source of competitive sustainability advantage. In the context of strategic management, two important aspects to consider for cooperation are (1) intentional and emergent planning stages, and (2) adaptability. First off, the phases of collaboration—which generally include context assessment, partner selection, joint strategy planning, and implementation—are the subject of a lot of research on cooperation. During these stages, careful planning is needed to evaluate the issue and its context considering the resources at hand, the qualities of probable partners, their quantity or degree of diversity, credibility, and any potential power imbalances. The execution of collaborative efforts can be impacted by contextually derived and imposed elements (Kristoffersen et al. 2020). As a result, a collaborative approach should include a thoughtful planning phase as well as flexibility to accommodate issues that occur throughout the process, which may come from the collaborators or the cooperation itself. The stages must be particular to the problems being worked on in the partnership and continuously adjusted to meet the demands of the partners (both the collective and each participating partner). Secondly, the necessity of customizing partnerships from the outset relates to the idea that greater flexibility is associated with higher collaborative success. These adjustments may involve changing contracts, decision-making processes, cooperative goals, or actor composition. Hence, rather than adhering to predetermined innovation or collaborative management trajectories, collaborators must be alert to emerging circumstances and actively start adjustments.

5.5.2 Open Innovation

Levels of Transparency, Obstacles, and Conflicts Open innovation study in the field of strategic management examines the knowledge management tactics and strategic choices needed to obtain, absorb, convert, or utilize information from beyond organizational boundaries through cooperative innovation. Open innovation study broadly looks at how businesses expand knowledge flows in and out of organizations to accelerate internal innovation, competitiveness, and performance. Investigations into "coupled" innovation—in which businesses create and commercialize discoveries together—are crucial to collaborative COI (Koop et al. 2021). Research on open innovation and the circular economy both particularly benefit from an understanding of the relationships and synergies between the various business models and the teamwork needed to keep things running smoothly throughout the course of product life cycles. According to open innovation research, there are two main things that businesses should consider before partnering: (1) the project's structure and degree of transparency, and (2) any possible conflicts or difficulties. Different project structures with varying degrees of transparency can emerge from strategic choices in collaborative innovation. The flexibility of collaborative innovation and the recognition that the organizational procedures, strategy, and capabilities of the business can determine the "appropriate" circumstances for choosing various collaborative structures.

5.5.3 *Circular Oriented Innovation Types- COI*

Its an all-encompassing word that encourages an innovative perspective that extends beyond the confines of a particular organization to modify or develop new systems. Businesses must pay more attention to how they adapt existing systems or develop new ones. Innovation must be integrated at all levels—product, process, organizational, and market. Across these similar kinds, collaborative innovation has a noteworthy and beneficial effect. We concentrate on these kinds of innovation as we are interested in company-to-company collaboration in innovation. In the context of circularity, product innovation entails improving the product's use-phase or recoverability through redesign, new product developments (NPD), and related services such as product-service combinations. This involves implementing principles and strategies of Circular Economy (CE) product design, such as reducing or replacing hazardous materials, creating long-lasting products with modular structures for repair and refurbishment, and integrating recovered materials. Process innovation, closely linked to product innovation, involves enhancing the use of renewable inputs or recoverability of outputs through new technologies or production techniques. Organizational innovation entails strategic decisions that introduce new arrangements or perspectives on issues like resource scarcity, supply chain, or environmental management. This may represent the framework of new company-to-company collaborative procedures inside collaborative COI (Kusiak 2023). Redesigning the organization's innovation process through the integration of new innovation concepts, enhanced reflexivity, and contact with outside actors is a kind of organizational innovation that is particularly pertinent to COI. While operation and organizational innovations focus on how to manufacture such items or ancillary services, product and service innovations are typically the focus of cooperation throughout the design and development phase. They also emphasize how organizational and process changes are essential prerequisites for creating new goods and services. This includes incorporating techniques for the CE business model, including service over ownership to increase the value of the product or residual value through recovery efforts. On the other hand, this makes innovation efforts more difficult. Lastly, there are two types of innovation: radical and gradual. The important distinction is whether the innovation activity seeks to produce completely new, non-contextualized processes, goods, services, technologies, and business models, or whether it alters existing ones.

It is currently unknown, nevertheless, how cooperative structures and procedures promote circularity for diverse goals among diverse stakeholders. For instance, it's not apparent what kind and degree of cooperation is needed to slow down and close resource cycles. It is important to remember that cooperation for circularity necessitates a variety of players both inside and across supply chains and industries, including governmental organizations, academic institutions, and other outside parties. As previously indicated, comprehending the significance of multi-actor, multi-level collaboration is essential for making the shift from a linear economy to a circular economy. However, there are still unanswered concerns concerning the management of multiple connections and the cooperation of diverse actors and stakeholders. For instance, identifying the potential to advance circularity and eventually making the transition to a CE requires cooperation between organizations and the government.

5.6 Developing a Circular Supply Chain Business Model Conceptually

In order to understand the origins of the concept of circular economy (CE) and how it is applied in real-world settings, it is necessary to take into account a number of related ideas, frameworks, and research areas, including regenerative design, industrial ecology, performance economy, blue economy, and cradle-to-cradle. It has been challenging to achieve a cross-disciplinary understanding of the CE notion since it has roots in a wide range of historical, economic, and ecological fields. As a result, diverse definitions, scopes, and CE narratives have been found in study and practice. Researchers believe that although the concept of CE is crucial for both academia and practice, there isn't much agreement on what it means. A few studies have been conducted to compile the most crucial concepts and conceptualizations in order to provide transparency to this still-developing field of study.

5.7 The Supply Chain Procedure in CBMs

Ostermann et al.'s study, "Business Model Innovation for Circular Economy in Fashion Industry: A Startups' Perspective," explores essential components of startups' Business Model Innovation for Circular Economy (BMI4CE) in the fashion sector (Ostermann et al. 2021). Motke et al.'s paper, "Circular Business Model Innovation and Its Relationship With Business Performance in Brazilian Industrial Chemical Companies," investigates the link between circular business model innovation and corporate performance in the Brazilian industrial chemical sector (Motke et al. 2022). Montag et al. develop a framework in "Paving the Way for Circular Supply Chains: Conceptualization of a Circular Supply Chain Maturity Framework" to analyze the degree of maturity of circular economy adoption in supply chains (Montag et al. 2021). Finally, Koop et al.'s paper, "Circular Business Models for Remanufacturing in the Electric Bicycle Industry," explores various circular business models combining additive manufacturing and remanufacturing in the electric bicycle industry (Koop et al. 2021).

5.8 Round Business Plans

A political objective of late has been the circular economy, especially in light of the growing costs of resources and the effects of climate change. But rather than being praised, the concept is more frequently questioned and condemned. In a circular economy, waste is repurposed into valuable resources through recycling and reuse, fostering the development of five key models: circular supply chains, recovery and recycling, product life cycle extension, shared platforms, and product as a service. Embracing these models requires businesses to shift away from linear thinking and adopt new business paradigms, offering benefits such as differentiation, cost reduction, new revenue streams, risk mitigation, and reduced reliance on virgin resources. While there's room for improvement, the adoption of these circular business models has surged in the past decade, initially driven by startups but now embraced by large multinational corporations as well. In order to execute sustainable business models, a number of supply chain participants must modify their procedures, practices, and attitudes while promoting the circular economy story. Customer segments, value propositions, channels, customer connections, income streams, critical resources, critical activities, critical partnerships, and cost structure are the nine components that make up a business model. Numerous combinations of these basic pieces enable the development of original business concepts.

5.9 Different Ways Within Business for Incorporating Circular Economy

There are four different ways for conducting business as shown in Fig. 5.2 for benefiting from the circular economy (Kusiak 2023).

5.9.1 Inner Circle

First, maintaining a product's viability and functionality for as long as possible ideally under the original owner or user—is the goal of the "power of the inner circle." The idea pertains to the maintenance loop and making sure that devices may be kept in good working order with simple upkeep and repairs, as well as potentially with handy upgrades. Developing supporting business structures and product designs are essential to seizing this potential. The most potential for profit is found in the inner ring of reuse and maintenance. In general, the savings should be greater in the inherent costs in terms of material, labour, energy, capital, and the accompanying issues of externalities, such as greenhouse gas emissions, water, or poisonous chemicals, the Ellen MacArthur Foundation notes.

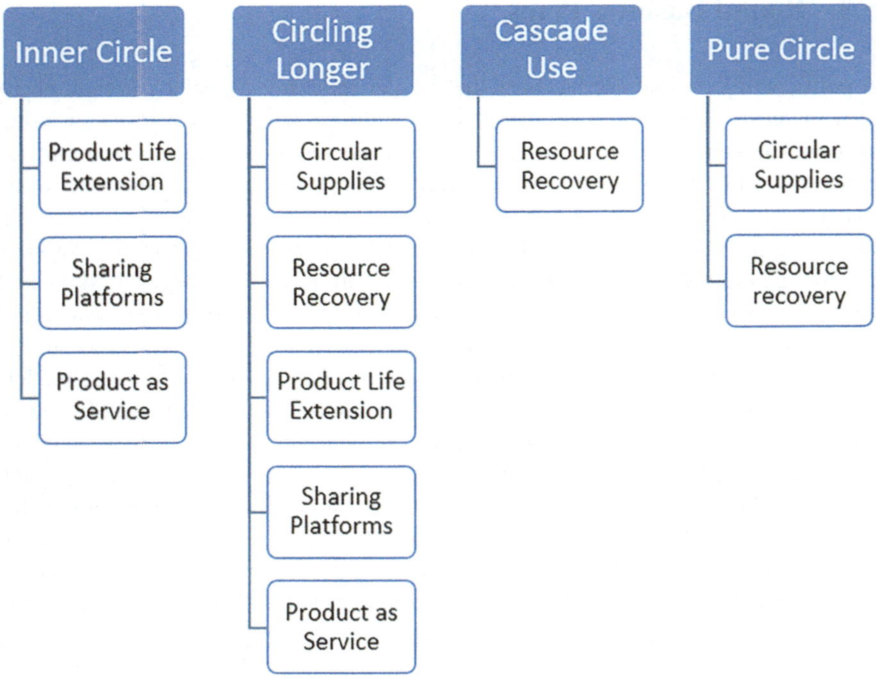

Fig. 5.2 Different circularity ways

5.9.2 Circling Longer

Second, maintaining products in as many successive cycles as feasible and extending the duration of each cycle is referred to as the "power of circling longer." In order to ensure a long life with the product's initial user, manufacturers of durable goods, like cars, could strive to offer things like high product quality, simple repairs and upgrades, or even service plans that guarantee the products stay in perfect condition. Step two can have a lengthy lifespan with additional customers, in which case the offerings are also relevant. Using the product as spare components to maintain the viability of other items might be the third step. As an alternative, parts from the previous goods may be incorporated into new ones of the same or different sort. The model that follows has a strong connection to the third value creation principle—cascaded use—which is discussed subsequently.

5.9.3 *The Strength of Using Cascades*

The third value-generating element that companies should consider incorporating into their circular business models is the "power of cascaded use." Diversifying the reuse of resources and products both within and between sectors is the goal. Clothes may be used as second-hand clothes in the garment business, upholstery in the furniture industry, and finally as part of an insulating material for building, which makes textiles an excellent example of cascaded usage. Reused goods and materials save the participating firms money on raw resources by replacing the influx of virgin material in each instance.

5.9.4 *The Pure Circle*

Lastly, the "power of pure circles" emphasizes the significance of clean material steams, since they are essential to preserving the materials' quality throughout a long series of cycles. One well-known eco-design approach that has been championed at least since the 1990s is the application of cleaner or purer materials. Still, the idea is as pertinent and significant as ever.

5.10 Circular Models That Work

Compared to the body of the circular economy, the idea of the circular business model is far more recent. The phrase initially surfaced in a 2006 paper by Schwager and Moser that examined several business model varieties for the generation of circular value. Seven years later, the circular business model idea reappeared at the same time as the World Economic Forum and the Ellen MacArthur Foundation began to promote the circular economy idea more widely (Schwager et al. 2017). The business model serves as a solution to the incentive alignment problems that account for the sluggish adoption of quickly developing technology that would facilitate the recycling of a wide variety of goods and materials. In a similar vein, business models are essential for the "long-term flourishing of the circular economy." In his view, firms ought to view the circular economy as a novel means of generating profits, rather than as a means of fortifying their operations against fluctuations in the cost of raw materials.

The company, a major global producer of air separation units, traditionally focused on manufacturing and distributing these systems for various industries (Tiwari et al. 2023). However, it has now diversified its approach by offering integrated technical equipment and industrial gases as part of complete product-service solutions (PSS). This expansion includes three additional PSS business models alongside its conventional product-based strategy.

5.10.1 PSS That Is Focused on Products

Apart from selling air separation product units, the company also provides supplementary market offerings for technical services such as installation, maintenance, and repair. This is the most commonly utilized PSS business model by the corporation. It follows a product-oriented PSS approach where customers acquire ownership of the items, and technical services are integrated into the initial sales package.

5.10.2 Use-Focused PSS

Product leasing and technical assistance Customers can hire the organization to lease a whole engineering system, or air separation units, for a certain number of years at a time. Customers with particular financial needs—such as those who lack the funds to purchase equipment or undertake construction projects—are the primary target audience for leasing contracts. Since clients only pay for the use of the items and services, the business maintains ownership of the products, forming a usage-oriented PSS.

5.10.3 PSS with a Focus on Results

In addition to offering gas generators, the company provides "industrial gas" services. In this result-oriented PSS model, the company owns the gas generators, and clients pay for the gases they use. The company initiated its first industrial gas project in 2003, with a significant increase in such projects occurring after its public listing in 2010. The company has witnessed a sharp surge in industrial gas projects in recent years, with over 38 gas centers constructed in Chinese industrial parks in 2015 alone. These gas centers produce various industrial gases, including O2, N2, CO2, H2, rare gases like Ar and He, and specialty gases. Gas is supplied to customers in bottles, liquid form (via cold air separation of liquid), gasification, and pipeline form for industrial parks (Kusiak 2023).

5.11 Company Model's Supply Chain for Circularity

The interviewees were asked to explain the supply chains associated with each business model in order to gain an understanding of how these various business models affected supply chain operations. The supply chains for the three different PSS business model types and the conventional product-based model which include producing and selling products are as follows and every company model's supply chain operation's circularity is examined below.

5.11.1 Conventional Product-Based Model

This model follows a linear supply chain pattern known as "take, make, and dispose," encompassing design, acquisition, manufacturing, distribution, consumption, and disposal stages. The company purchases components and raw materials from suppliers, manufactures products, distributes them to its customers, who dispose of the goods after use. Information and supplies are not being returned to the firm. Remanufacturing and reuse occur at a relatively low rate. Sometimes simply minor damaged pieces lead to the entire product being scrapped.

5.11.2 PSS Focused on Products

According to this business plan, the corporation offers technical support for the goods while they are being used. In contrast to the supply chain of a product-based business model, a product-oriented PSS extends the product's lifespan through regular maintenance and repair services. Both the company and consumers benefit from technical services as they reduce costs for customers compared to establishing their own service teams and increase service revenue for the company. Manufacturers can offer more expert services because they have a deeper understanding of their products than consumers or third-party businesses, providing an advantage in delivering technical services.

5.11.3 Practical PSS

In its use-oriented PSS, the company provides technical services and leases equipment to customers. Unlike the previous models, leasing projects in this category allow for more cycles of reuse, recovery, remanufacturing, and recycling of the spent gas generators. The rationale is that the business maintains ownership of the items and has an incentive to maximize the value derived from end-of-life products. The more time consumers spend using a product, the more money the business may make by leasing it.

5.11.4 A PSS Focused on Results

Instead of selling gas generators, the company sells industrial gases under this business model. Rather of producing and supplying gas generators to clients, the firm produced gas centers in industrial parks near clients. Gas generators were really purchased by the corporation, and the generated gasses were supplied to different industrial park clients. In this model, the company, being the actual user, has greater control over the usage phase, enhancing its ability to implement maintenance, repair, reuse, recovery, remanufacturing, and recycling processes compared to the previous three business models.

5.11.5 Key Forms of Value Generation

The company's product-based business model strictly adheres to linear supply chain principles, lacking intrinsic circularity. In contrast, all three PSS business models exhibit varying degrees of circularity, with use-oriented PSS demonstrating the highest influence on extending inner circles and increasing their duration. This suggests that use-oriented PSS has longer and more frequent inner circles compared to product-oriented PSS. Additionally, the result-oriented PSS may generate value through cascaded usage, leveraging co-products and by-products in different value chains, alongside inner and longer circles. We can lead the market need if we can take advantage of this opportunity and investigate new gas uses. This form of circularity, particularly pure reuse of materials, is unique to result-oriented PSS, as only this model allows manufacturers to manage and derive value from the byproducts and co-products of their primary goods through cascading usage (Schwager et al. 2017). Respondents demonstrated a stronger inclination towards achieving pure reuse of materials under result-oriented business models, even though pure circles may not seem significant to PSS models. This difference in circularity between the business models is primarily attributed to product ownership dynamics. As the company retains ownership in use and result-oriented PSS models and exercises significant control over products throughout their life cycles, it is more incentivized to extract value from the air separation units and their associated products (e.g., O2, H2, Ar, He, Xe, and Ke). Consequently, the company is motivated and obligated to reduce the environmental impact of these products, especially when such efforts positively affect their financial value.

5.12 Conclusion

The implementation of smart manufacturing practices, driven by the principles of the circular economy, marks a significant shift in how businesses approach production, resource management, and sustainability. The transition from linear to circular models offers industries the opportunity to reduce their environmental impact, while also optimizing resource use and reducing costs. The circular economy model not only minimizes waste by focusing on recycling and remanufacturing but also ensures that products are designed for longevity through reparability and durability. This fosters a system where resources are continuously cycled back into production, minimizing the need for raw material extraction and reducing the environmental burden.

Furthermore, the adoption of circular supply chain models allows companies to maximize product efficiency at every stage of the production cycle. By incorporating strategies such as closed-loop systems and collaborative consumption models, businesses can extend product life-cycles, reduce waste, and promote resource sharing. These approaches contribute to greater economic resilience and align with the growing demand for environmentally responsible production methods. However, the widespread adoption of these practices still faces challenges, including the need for more robust infrastructure, cross-sector partnerships, and deeper understanding of circular supply chain dynamics. In conclusion, smart manufacturing, underpinned by circular economy principles, offers a sustainable path forward, balancing economic growth with environmental stewardship.

References

Amjad MS, Diaz-Elsayed N (2024) Smart and sustainable urban manufacturing for a circular economy. Environ Dev Sustain:1–27

Atif S (2023) Analysing the alignment between circular economy and industry 4.0 nexus with industry 5.0 era: an integrative systematic literature review. Sustain Dev 31(4):2155–2175

Blunck E, Werthmann H (2017) Industry 4.0–an opportunity to realize sustainable manufacturing and its potential for a circular economy. In: DIEM: Dubrovnik International Economic Meeting. Vol. 3. No. 1. Sveučilište u Dubrovniku

Bressanelli G et al (2022a) Circular economy in the digital age. Sustain For 14(9):5565

Bressanelli G et al (2022b) Towards the smart circular economy paradigm: a definition, conceptualization, and research agenda. Sustain For 14(9):4960

Chowdhury NR et al (2023) Implementing smart waste management system for a sustainable circular economy in the textile industry. Int J Prod Econ 262:108876

Ciliberto C et al (2021) Enabling the Circular Economy transition: a sustainable lean manufacturing recipe for Industry 4.0. Bus Strateg Environ 30(7):3255–3272

Cioffi R et al (2020) Smart manufacturing systems and applied industrial technologies for a sustainable industry: a systematic literature review. Appl Sci 10(8):2897

Ćwiklicki M, Wojnarowska M (2020) Circular economy and industry 4.0: one-way or two-way relationships? Eng Econ 31(4):387–397

Ghaithan AM et al (2023) Integrated impact of circular economy, industry 4.0, and lean manufacturing on sustainability performance of manufacturing firms. Int J Environ Res Public Health 20(6):5119

Gómez AMM, González FA, Bárcena MM (2018) Smart eco-industrial parks: a circular economy implementation based on industrial metabolism. Resour Conserv Recycl 135:58–69

Han Y et al (2023) Exploring how digital technologies enable a circular economy of products. Sustain For 15(3):2067

Koop C et al (2021) Circular business models for remanufacturing in the electric bicycle industry. Front Sustain 2:785036

Kristoffersen E et al (2020) The smart circular economy: a digital-enabled circular strategies framework for manufacturing companies. J Bus Res 120:241–261

Kusiak A (2023) Smart manufacturing. In: Springer handbook of automation. Springer International Publishing, Cham, pp 973–985

Montag L, Klünder T, Steven M (2021) Paving the way for circular supply chains: conceptualization of a circular supply chain maturity framework. Frontiers in Sustainability 2:781978

Motke FD et al (2022) Circular business model innovation and its relationship with business performance in Brazilian industrial chemical companies. Frontiers in Sustainability 2:766696

Ostermann CM, da Silva L, Nascimento, and Aurora Carneiro Zen. (2021) Business model innovation for circular economy in fashion industry: a startups' perspective. Frontiers in sustainability 2:766614

Schwager P, Dunjic B, Kaltenegger I (2017) Success and failure of the Chemical Leasing model in addressing sustainability challenges: Evidence from practice. Current Opinion in Green and Sustainable Chemistry 8:14–17

Tiwari S, Bahuguna PC, Srivastava R (2023) Smart manufacturing and sustainability: a bibliometric analysis. BIJ 30(9):3281–3301

Chapter 6
Understanding the Risk Associated in Implementing Circular Economy

Abstract The circular economy (CE) aims to move businesses from a linear model of "take-make-dispose" to a circular one, where products and resources are continuously reused, remanufactured, or recycled. This transition presents significant opportunities for sustainability and innovation but also introduces various risks. Implementing a circular economy strategy requires a clear understanding of the organization's goals, readiness, and the associated risks, including economic, environmental, social, cultural, and regulatory challenges. Companies must align their business models, supply chains, and stakeholder relationships with circular economy principles to ensure long-term success. The circular economy involves new financing methods, regulatory adjustments, and the promotion of sustainable technologies. Risk management becomes crucial in this complex ecosystem, with strategies that balance environmental, social, and financial outcomes. Adopting circular innovation, modifying business models, and enhancing internal capabilities are key steps to successfully mitigating risks and advancing the circular economy. Legal considerations and strong partnerships are essential for a smooth transition.

Keywords Circular economy · Risk management · Sustainable innovation · Stakeholder engagement · Regulatory compliance · Supply chain optimization

6.1 Introduction

Establishing what the circular economy means the company and its unique objectives and aspirations is the first step towards circular economy (CE). For instance, to decrease the amount of waste produced, utilize more renewable energy, make your products last longer, or turn waste materials into new sources of income. The objectives for the circular economy should be clearly communicated to your stakeholders and should be in line with your entire corporate social responsibility (CSR) strategy and vision. The next stage is to determine how prepared the company is to put circular economy ideas and practices into effect. To examine present performance and any gaps in areas like product design, material sourcing, production processes,

J. Kandasamy et al., *Circular Supply Chain Management*, Circular Economy and Sustainability, https://doi.org/10.1007/978-3-031-81750-2_6

distribution channels, customer involvement, and end-of-life management, the system can utilize a maturity model or a self-assessment tool. The circular economy's ethical, legal, and regulatory ramifications for your company's operations and compliance should also be taken into account. The road toward understanding CE as a long-term strategic objective for competitiveness in international markets suggests a profound capacity for risk management associated with the modification of the corporate operation. In order to remain successful in international markets, a company must engage in a complicated decision-making process that is guided by its goal, vision, and core values (AlMashaqbeh and Munive-Hernandez 2023). This process is known as strategy. The growing significance of the relationship between strategy and corporate performance necessitates a thorough analysis of strategic risk, or the chance that certain events may transpire and have an impact on the accomplishment of associated business objectives and CE plans. Stated differently, the transition from a linear to a circular vision necessitates a careful assessment of the risks associated with the future and the company's actions within it. A company's capacity to meet its objectives and create value for its stakeholders and for itself may be hampered by strategic risk. Thus, in order to minimize possible roadblocks to success and optimize organizational performance, executives adopt strategic risk management, which is the process of identifying risks that originate from both the inside and the outside, as well as their causes and consequences, and then taking appropriate action to mitigate them. In a dynamic notion of strategy, risk also includes opportunities that may present themselves as a result of uncertainty (Blomsma et al. 2023). As a result, circumstances that hinder the accomplishment of some goals may make it difficult to provide value for other goals. Because the system is the repository of the risk-aware corporate culture and has a comprehensive perspective on the whole business and its aims for producing value over the long term, the integrated risk-focused approach to strategy requires the assistance of the system.

6.2 Risk Involved in Implementing (Awan and Sroufe 2022)

Figure 6.1 indicates the various risk associated with implementing the circular economy concept in an organization.

6.2.1 Economic Risk

Getting the money or financing required to put a circular economy plan into action is the fourth hurdle. Compared to the linear economy model, you could have to deal with greater upfront expenditures, longer payback times, or worse returns on investment. Additionally, you can have trouble getting funding from conventional sources like banks or investors who might not be aware of or appreciate the circular

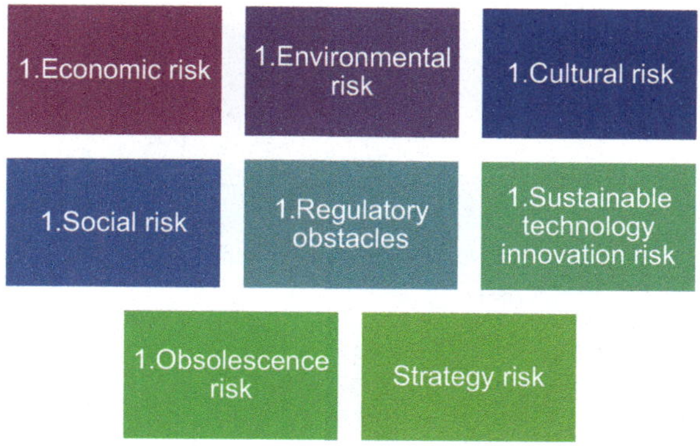

Fig. 6.1 Risks associated with implementing the circular economy concept

economy. In order to overcome this obstacle, you must present the long-term advantages and business case for a circular economy (Agyemang et al. 2019). You also need to look at novel or alternate funding options, such as grants, subsidies, crowd sourcing, or green bonds. The following figure (Fig. 6.2) showcases the list of economic associated risks.

1. **Competition Risk**

   Competition risk refers to the potential for a business to lose its market share or profitability due to the actions of competitors. In a highly competitive market, companies may face price wars, innovation challenges, or decreased demand for their products or services as competitors offer better alternatives. Firms need to continuously innovate, improve customer experience, and differentiate themselves to mitigate competition risk.
2. **Interest Rate Risk**

   Interest rate risk arises from fluctuations in interest rates that can affect a company's borrowing costs and investment returns. If interest rates rise, businesses with significant debt might face higher interest expenses, reducing profitability. Conversely, declining interest rates can lower borrowing costs but may reduce the returns on certain investments, especially those tied to interest-sensitive assets.
3. **Exchange Rate Risk**

   Exchange rate risk, also known as currency risk, occurs when changes in foreign exchange rates impact the value of cross-border transactions, investments, or earnings. Companies operating in multiple countries or trading in foreign currencies are exposed to this risk, as unfavourable currency movements can erode profits. Hedging strategies like forward contracts or options can help manage exchange rate fluctuations.
4. **Supplier Price Risk**

   Supplier price risk is the risk that changes in the cost of goods or services provided by suppliers will impact a company's profitability. Factors such as inflation, shortages of raw materials, or geopolitical events can lead to supplier price hikes, putting pressure on a company's margins. Building strong relation-

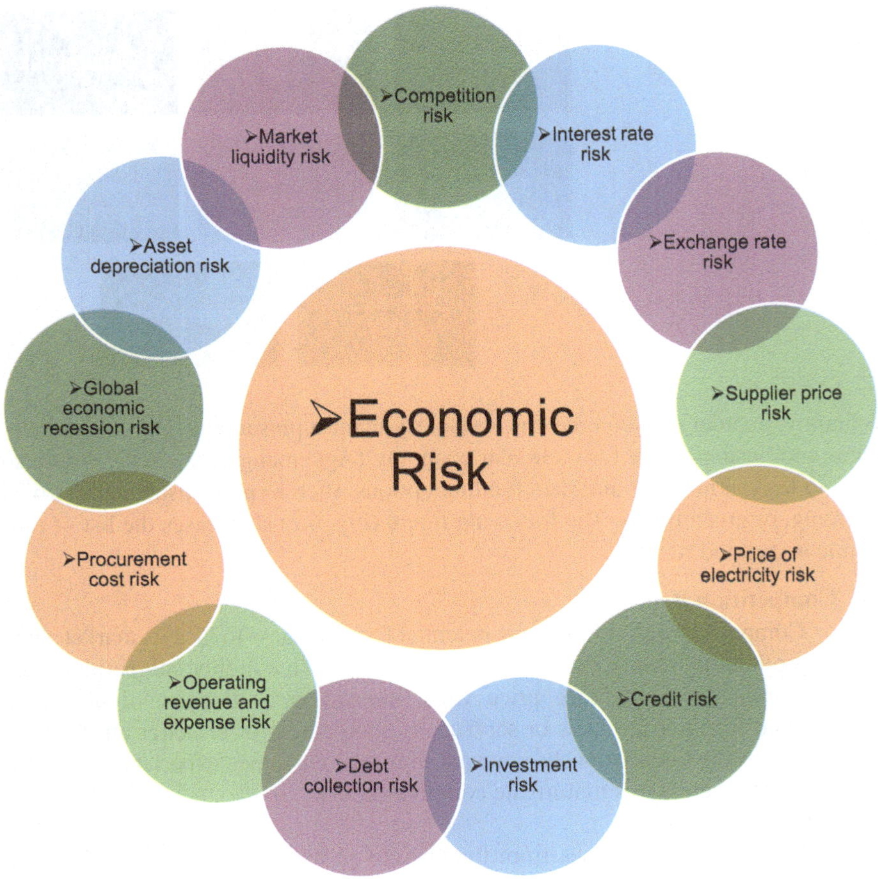

Fig. 6.2 Economic risks

ships with suppliers, negotiating long-term contracts, or diversifying the supplier base are common strategies to manage this risk.

5. **Price of Electricity Risk**

   The price of electricity risk relates to the volatility in energy costs, which can significantly impact industries that rely heavily on electricity. Fluctuations in electricity prices can result from changes in demand, supply disruptions, regulatory changes, or shifts in energy markets. Companies can mitigate this risk by investing in energy efficiency, renewable energy sources, or securing fixed-rate energy contracts.

6. **Credit Risk**

   Credit risk is the potential for financial loss due to a borrower's inability to repay a loan or meet contractual obligations. Companies extend credit to customers, but if these customers default, it can result in significant losses. Firms manage credit risk by conducting thorough credit assessments, setting credit limits, and diversifying their customer base to reduce dependency on any single debtor.

7. **Investment Risk**
   Investment risk refers to the possibility that investments may not yield the expected returns, or worse, lead to financial losses. This can arise due to market volatility, poor management decisions, or external economic factors. Companies invest in various assets or business ventures, and they must weigh the potential risks and rewards, often diversifying their portfolios to minimize exposure.
8. **Debt Collection Risk**
   Debt collection risk is the likelihood that a company will face difficulties in recovering payments from customers or clients who owe them money. Poor credit control or deteriorating economic conditions can lead to higher incidences of bad debts. Effective credit control measures, prompt invoicing, and engaging professional debt collection services can help mitigate this risk.
9. **Operating Revenue and Expense Risk**
   Operating revenue and expense risk refers to the variability in a company's income and costs due to market conditions, operational inefficiencies, or unexpected events. Companies may experience fluctuations in their revenue streams or incur unexpected operating costs that impact profitability. Effective budgeting, cost control measures, and diversifying revenue streams can reduce this risk.
10. **Procurement Cost Risk**
    Procurement cost risk is the risk that the costs associated with acquiring goods or services will increase unexpectedly. This can be due to supplier price hikes, changes in market conditions, or disruptions in the supply chain. Companies can mitigate this risk by negotiating fixed-cost contracts, diversifying suppliers, and maintaining efficient procurement processes.
11. **Global Economic Recession Risk**
    Global economic recession risk refers to the potential impact of a global downturn on a company's performance. A recession can reduce demand for goods and services, leading to decreased sales, profitability, and cash flow issues. Companies can prepare for such risks by building cash reserves, diversifying markets, and focusing on cost-efficiency.
12. **Asset Depreciation Risk**
    Asset depreciation risk is the risk that the value of a company's assets will decrease over time, reducing its overall net worth. This can occur due to wear and tear, technological obsolescence, or changes in market conditions. Managing depreciation risk involves regularly maintaining assets, investing in newer technology, and carefully monitoring the lifespan of key equipment or properties.
13. **Market Liquidity Risk**
    Market liquidity risk occurs when a company is unable to sell assets quickly enough to meet its financial obligations without significantly lowering the asset's price. In times of financial distress or market turbulence, liquidity can dry up, forcing companies to sell assets at a discount. Maintaining a diversified

portfolio of liquid assets and access to credit lines can help manage liquidity risk.

6.2.2 Environment Risk

Evaluating and reducing the possible adverse environmental effects of a circular economy represents the fifth challenge (Choudhary and Kumar 2024). A circular economy has the potential to save emissions and waste, but it may also introduce unanticipated hazards or new environmental challenges. For instance, you could use more water or energy, produce new kinds of pollution, or create new risks. To overcome this obstacle, you must evaluate the risks and life cycle of your circular economy initiatives and take action to stop or lessen any negative consequences.

1. GHG emissions (NOx, CO_2 and SO_2) risk
2. Environmental regulations
3. Industrial water reuse ratio risk
4. Recycling of treated water risk
5. Solid waste risk in thermal power plants
6. Waste handling risk
7. Lost time due to injuries risk
8. Accident fatalities per energy produced risk (Severe accidents risks)
9. Human toxicity potential expresses risk (ex. Polychlorinated Biphenyls (PCBs))
10. Noise impact caused by energy system risk
11. Bad Odors risk
12. Risk of mortality due to normal operation (reduced life expectancy in years of life lost/GWh)
13. Soil pollution risk

6.2.3 Cultural Risk

Changing the attitudes and actions of your stakeholders—such as suppliers, workers, investors, and customers—presents another difficulty. They could not perceive the benefits or viability of a circular economy since they are accustomed to the take-make-dispose linear economy paradigm (Dulia et al. 2021). In order to overcome this obstacle, you must convince people of the advantages and effects of a circular economy while also offering encouragement and assistance in its adoption. Additionally, you must promote an innovative and cooperative culture both inside and outside of your value chain.

6.2.4 Social Risk

Both the creation and loss of jobs are frequently discussed in relation to the social impacts of the circular economy. Although it is undeniable that having a job, a stable income, and safe working conditions are crucial, these factors do not fully capture social effects. The social implications of circularity do not yet seem to have a common definition, which makes measuring and taking into account them while implementing a circular economy particularly difficult (Dan et al. 2023). The risk associated with social aspects while implementing circularity is listed below.

1. Lack of motivation for staff risk
2. Lack of innovation risk
3. Lack of organizational learning capability risk
4. Poor relationship between parties' risk
5. Labor strikes risk
6. Social challenges risk (Poverty, substantial levels of inequalities, health challenge)
7. Behavioural aspect of employee's risk
8. Union/labour relations risk
9. Reputation risk (Negative Media Coverage)
10. Changing behaviour risk (Change in human behaviour)
11. Local community impacts risk

6.2.5 Regulatory Obstacles

Adopting a circular economy strategy might be difficult when it comes to following laws that were not created with it in mind. For instance, you can have trouble satisfying safety and quality requirements for recycled products or moving or reusing resources across international boundaries. Maintaining current knowledge of pertinent laws and regulations is essential to overcoming this obstacle (Dan et al. 2023). The system should also interact with regulators and lawmakers to push for more unified and helpful frameworks.

6.2.6 Sustainable Technology Innovation Risk

Identifying or creating the appropriate technology and solutions to support a circular economy is the third problem. For example, in order to create, produce, or utilize items more effectively and efficiently, you might need to invest in new machinery or procedures. It can also be necessary for you to use or develop platforms or systems that make resource recovery, leasing, or sharing easier (Gennari and Cassano 2020). In order to overcome this obstacle, you must work with other businesses or specialists who can offer or co-create the solutions you want, as well as do research and analysis on the innovations and best practices in your industry.

6.2.7 *Obsolescence Risk*

Overcoming early obsolescence, premature product failure will dramatically reduce overall ownership costs and increase convenience by removing the headache of returns and repairs. A fundamental change in our entire socioeconomic model is required for the circular economy to succeed. It requires that resources be purposefully allocated early in a product's design process to extend its useful life and that, when it does become practically obsolete, it be easily disassembled into its constituent parts so that they can be recycled and repurposed (dos Santos Leite Gonella et al. 2024). It is the responsibility of the product's creator, manufacturer, distributor, and end user to consider such factors.

6.2.8 *Strategy Risk*

In order to balance the long-term view of social and environmental outcomes with the short-term pressures for financial performance, organizations involved in CE strategies should integrate risk management into the entire process, which includes the planning, the assessment, and the achievement of strategies (Halkos and Aslanidis 2024). When potential courses of action for the indicated risk response are discovered, the loop closes itself. When a strategy fails, it is necessary to go back and identify any gaps and associated causes. The following have been discussed in detail

1. Strategies for the Circular Economy and Governance Risk.

   Governance risk in the context of a circular economy (CE) refers to the challenges organizations face in aligning their corporate structure, policies, and decision-making processes with circular principles.
2. Strategies for the Circular Economy and the Risk of Stakeholder Relationship Management

   The risk of stakeholder relationship management in the circular economy refers to the potential challenges businesses face when engaging with key stakeholders such as suppliers, customers, investors, regulators, and the local community. In a circular economy, businesses are dependent on stakeholders for support in areas like resource sourcing, product design, recycling, and waste management. Poor stakeholder management can lead to misunderstandings, delays in implementing circular initiatives, or resistance to change.

3. Strategies for the Circular Economy and Innovation Risk

   Innovation risk in the circular economy context refers to the potential challenges and uncertainties businesses face when developing new technologies, products, or business models to support circularity. Transitioning to a circular economy often requires organizations to rethink product design, manufacturing processes, and business models. These innovative changes involve a level of risk, including failure to achieve desired outcomes, high development costs, or a lack of market acceptance.

6.2.8.1 Strategies for the Circular Economy and Governance Risk

Corporate boards that place a high priority on innovation, sustainability, and resilience are the ones who create circular economy plans. Board commitment drives the planning stage, which formalizes the strategy in accordance with the mission, vision, and corporate core values. Related actions and the boundaries within strategies should be issued to ensure coherence between the entity's goals, decisions, actions, and stakeholder communication (dos Santos Leite Gonella et al. 2024). In order to lower the risk associated with a strategy that is not in line with the purpose, vision, and core values, it is crucial to take a comprehensive approach when considering CE strategies around the sustainability value drivers (improving resource efficiency, extending lifetime, and completing the loop). The capacity of the board to develop CE strategies while taking the board's structure, accountability, and responsibilities into consideration is connected to this governance risk. A lot of research has been done on the relationship between board structure and a company's level of sustainability focus. Stakeholder protection techniques include separating the CEO and chairman roles, creating committees focused on sustainability problems, and limiting the number of non-executive and independent members (Halkos and Aslanidis 2024). These measures help to lower the likelihood that short-term orientation would trump long-term wealth generation. To ensure that stakeholder needs are appropriately taken into account, the board must assess its own nomination, compensation, and composition policies. International regulators, who are pressing for global competitiveness while protecting the environment and granting new rights to consumers, are also emphasizing the alignment between corporate strategies and global expectations (Souza Piao et al. 2024). These regulators are targeting specific industries, such as electronics, textiles, plastics, and packaging, with their actions.

Furthermore, due of the technical complexity of circular projects and the difficulty of evaluating them in terms of future benefits, CE plans call for the talents and sensitivity of specific directors. Potential trade-offs between alternate circular and non-circular tactics may present themselves in this situation, underscoring the board's need to accurately determine the balance between present efforts and future success. Moreover, the board has the responsibility and accountability of clearly communicating expectations to stakeholders and fostering enduring relationships with them. The full value chain must be involved for the CE concept to be really implemented in business strategy; in order to close the loop, all parties—from regulators and resource suppliers to ultimate consumers—must play a crucial role. Adopting a circular model necessitates a drastic shift in how things are thought about, done, and consumed, including products and services. Stated differently, the board assumes the task of modifying the conventional business model or implementing an entirely new one (Kottmeyer 2021). A company's business model outlines its operations and converts strategy into value. It illustrates the value proposition, the network to distribute the value produced to markets, and the resources as assets and key capabilities. The Board can offer a framework for identifying critical risks by using the formal definition of CE business models, keeping in mind that CE necessitates the consideration of outcomes rather than outputs and a more intricate system of players participating in the strategy's achievement. Circular business models are deemed inappropriate when boards undervalue these developments and have a short-term perspective.

6.2.8.2 Strategies for the Circular Economy and the Risk of Stakeholder Relationship Management

The board looks to discover and comprehend the possible risks associated with each alternative strategy while assessing it. Strategies are often assessed based on the performance they can attain, taking into account their essential assumptions and existing and future relationships with internal and external stakeholders. The idea of corporate performance serves as the foundation for the whole strategy assessment process. Although the term "CE" lacks a widely agreed definition, it is acknowledged that it is a component of the sustainable development idea. Therefore, the board's review of CE initiatives is predicated on this broad definition of success for many stakeholders, as well as the company's obligation to mitigate the possibility that the whole assessment procedure is misunderstood (Obiuto et al. 2024). The capacity of CE strategies to meet stakeholder expectations while taking the established level of acceptable risk into account is referred to as acceptability. Businesses build a range of relationships with several stakeholder groups, each of which expresses or implies a particular set of expectations and experiences under performance. By placing stakeholders on a map and taking into account the significance of the issues surrounding CE projects for both the company and them in accordance with the notion of materiality, it is possible to find a balance between these expectations and highlight the priorities from a comparative perspective. The risk of placing undue emphasis on certain stakeholder groups or short-term expectations must be considered in this stage of the strategy assessment process. Building strong relationships with stakeholders and evaluating associated risks are based on a framework that reflects the internal capabilities and the external environment. Both external opportunities and internal strengths can be advantageous to many relationships, but many also suffer from internal shortcomings and external limits.

Companies must develop strategies in this environment that, by utilizing their unique resources, ensure a long-term sustainable competitive edge. Following an evaluation of CE's fitness for the external environment and the stakeholders that inhabit it, a thorough examination of the strategy's internal sustainability is conducted. Internal resources are defined as tangible assets such as income and property, as well as intangible assets like people and their relationships that are integrated into the value chain (Patra et al. 2024). These resources are arranged within the organization to maximize their synergies, taking into account the long time horizon that characterizes the return on CE investments and the challenge of estimating social and environmental outcomes.

6.2.8.3 Strategies for the Circular Economy and Innovation Risk

Internal control systems that guarantee the strategic objectives are managed in accordance with a comprehensive corporate perspective are necessary for the firm to successfully implement the CE goals. Investing in innovation to find alternatives to the linear model—which provide definite positive cash flows in the future and real negative cash flows in the present—is, in fact, what defines the transition to circular business models. Because of this, a CE strategy that the board deems feasible must be translated into practical goals while keeping in mind that there is a delay between present efforts and potential rewards. The financial sector plays a crucial role in assisting businesses in their innovation process towards the circular economy. This is achieved through long-term investment strategies that incorporate financial, social, environmental, and governance analyses in the evaluation of organizations (Shooshtarian et al. 2024).

A proper time frame for evaluating the success of investments and a suitable selection of relevant performance measures are mandated by the time gap and the ESG (environmental, social, and governance) evaluation criteria. There is a genuine chance that present pressures, brought on by unforeseen circumstances or catastrophic occurrences like sanitary emergence, may sever a sustainable course. As a result, risk management is also considered while implementing CE plans. The corporate risk is the focal point of the present internal control system, which emphasizes the significance of integration with enterprise risk management. As a result, after the board has determined which strategies are pertinent for a circular way of doing business, the next step is to confirm that the internal control system is appropriate. This is done by beginning with the management control to specify roles, outcomes, and associated risks that could have an impact on processes. The spread of circular values and employee motivation to follow them, in particular, are the cornerstones of CE-focused management control. Goals from strategic planning are also incorporated into targets, action plans, and performance measurements, with a greater emphasis on the product life cycle (from cradle to cradle) and new KPIs (Singha et al. 2024). Due to its longer time horizon and potential for short-term economic benefit, CE strategies necessitate modifications in the evaluation of operational decision-making processes and associated goals, with a focus on cost accounting and investment budgets.

Therefore, companies need to understand that implementing CE strategies without first adopting a new viewpoint on how to manage the business might increase the likelihood of a mismatch between new important success elements and traditional KPIs. When objectives are appropriately identified, they can also help with the move towards CE since their success validates the soundness of the approach. The efficiency of information systems aids in the board and management's evaluation of CE investments. Information flows between governing bodies and the company serve as the foundation for internal control procedures. The amount of data generated by new IT increases exponentially, as does the ease with which it can be processed and stored (Patra et al. 2024). In this situation, there is a chance that information overload will cause the organization and its stakeholders to make poor judgments. Finding useful and actionable information in a sea of seemingly unstructured data is a problem for risk managers. Therefore, maintaining a high standard of quality in internal information is crucial for the proper formulation, evaluation, and implementation of CE strategies. Similarly, maintaining a high standard of quality in external information is crucial for the growth and upkeep of positive relationships with stakeholders.

6.3 Determine and Analyze Your Risks (Shooshtarian et al. 2024)

Prior to starting any circular economy project, you should determine and assess any possible hazards. This covers hazards related to operations, finances, legality, reputation, the environment, and society, as well as risks that are external and internal. You may evaluate your risks and rank them based on likelihood and effect by using techniques like SWOT analysis, risk matrix, and scenario planning. By doing this, you may plan ahead and anticipate the possibilities and obstacles that your move to a circular economy may present.

6.3.1 Sync Your Objectives

Aligning your aims and objectives with those of your partners and stakeholders is another essential step in minimizing the risks associated with the circular economy. This entails making sure that your company's circular economy vision and strategy are understood, applied consistently, and shared with all key stakeholders, including suppliers, consumers, investors, regulators, and other relevant parties (Hassan et al. 2024). Your stakeholders and partners will collaborate, trust, and support you if you establish a shared understanding and commitment to your circular economy goals. You can also lower the likelihood of misalignment, disagreement, or opposition.

6.3.2 *Modify the Way You Do Business*

Your company model might need to change in order to conform to the new paradigm of value generation and delivery in order to effectively migrate to a circular economy. This might entail making adjustments to your supply chain, revenue model, product design, service offering, or customer interactions. For instance, you may go from owning assets to sharing or leasing them, from selling items to offering services, or from linear to circular supply chains. You may reduce resource consumption and waste generation while increasing value creation for consumers and stakeholders by making changes to your business strategy.

6.3.3 *Take Up Circular Innovation*

Adopting circular innovation, which is the act of creating new or improved solutions that promote circularity, is another strategy to handle the risks associated with the circular economy. This might involve innovation in systems, business models, procedures, or technology. For instance, you may leverage digital technologies like blockchain, IoT, and AI to provide intelligent resource and product tracking, monitoring, and optimization. Modularity, robustness, and biodegradability are examples of circular design concepts that you may apply to improve the lifetime and functioning of your products (Stylianou et al. 2024). You may reduce your negative effects on the environment and society while increasing your productivity, performance, and competitiveness by using circular innovation.

6.3.4 *Develop Your Abilities and Culture*

Lastly, you must improve your organization's culture and skill set in order to handle the risks associated with the circular economy. This entails acquiring the skills and knowledge necessary for the shift to a circular economy, such as resilience, creativity, systems thinking, and teamwork. You may accomplish this by offering your staff education, training, or mentorship programs in addition to promoting learning, trying new things, and getting feedback. A culture of circularity, or a way of thinking and doing that accepts circularity as a fundamental concept and value, is something else you must cultivate (Kottmeyer 2021). This may be achieved by clearly defining expectations, providing incentives and rewards for your staff, and highlighting your successes and effects related to the circular economy.

6.3.5 *Legal Perspective*

Companies that wish to participate in the transition to a circular economy must take certain legal factors into account. Circular policies will require a great deal of invention, funding, and teamwork to integrate into supply chains and brand identities. The following actions could be taken: funding innovation and research and development (R&D) to create components and materials that are easier to recycle or from which raw materials can be extracted more easily; or investing in business practices that guarantee or promote circularity. Increasing internal and external cooperation to attain the necessary level of innovation and service delivery. Encouraging creativity among employees (Singha et al. 2024). When suppliers are unable to provide goods that are sufficiently environmentally friendly or circular, old contracts should be terminated, and new agreements should be negotiated. Hiring these experts or appointing consultants to offer advice on circularity.

Businesses have a chance to innovate with regard to their goods and products thanks to the circular economy. This in turn may draw in new customers and workers who share a commitment to sustainability, especially from younger generations. Implementing circular policies, however, also carries the risk of materially altering or disrupting supply chains and generating liabilities; for instance, in the event that a supplier's contract termination is handled incorrectly or advertising claims regarding a product's circular quality prove to be unfounded. When establishing new supplier partnerships and procedures, businesses should take careful consideration into account. This includes reaching clear agreements regarding who will hold the intellectual property (IP) resulting from these innovations. Here, confidentiality responsibilities, joint venture (JV) agreements, and research and development (R&D) cooperation agreements are all crucial (Souza Piao et al. 2024). So is a careful analysis of the competition law implications when R&D collaborations involve current or prospective rivals in any given market.

6.3.6 *Partnerships/Collaborations*

Collaborations can occur between governmental or academic organizations and corporations, or between corporate entities. They might be set up as one-time initiatives, continuing joint ventures, or extended partnerships. The parties in a research and innovation partnership agreement should address the wide variety of issues raised by these arrangements. Things to think about are as follows: How these new frameworks and resources are going to be created. What each partner will bring to the table in terms of resources, including lab facilities used. What expertise, information, and intellectual property will each partner provide to the project? Who will be entitled to use or own the know-how, data, and intellectual property (IP) in the final products or processes? The R&D's financial source. Whether or if secondments will be used to staff the projects, and if so, how access to trade secrets and shared information will be managed (Shooshtarian et al. 2024).

Regardless of whether the connection succeeds or fails, this will demand an awareness of what each participant contributes to the table and what each can retain or utilize going forward. The foreground and background intellectual property and know-how should be determined as early as feasible, if not before, to avoid any potential for relationship problems to develop. These days, data are also a primary outcome of most partnerships; ownership and access to this data should be thought through early on, in case the cooperation stops. Clarity about confidentiality is necessary, and before any activity or information sharing commences, non-disclosure and confidentiality agreements should be established. While maintaining secrecy is a major business goal, it is especially crucial when it comes to intellectual rights, since it safeguards any future patent applications that may otherwise be rejected for earlier disclosure.

6.3.7 Vigilance in Supply Chains

Companies will either need to contract for the supply of certified recovered goods from other sources, or remove essential elements for reuse from their products at the end of their life, in order to support resource recovery rules. For a company to advertise that it is 100% circular, every stage of the supply chain must be verified as circular. If this isn't feasible, companies will have to determine which connections in their value chain aren't circular by doing their due diligence (Obiuto et al. 2024). When vendors present themselves as circular, companies should make sure that appropriate indemnities and guarantees are in place to cover any future breaches and to offer remedies in the event that the initial promises turn out to be false.

6.3.8 Current Agreements

Materials supply contracts may eventually need to: Allow for variable purchase volumes rather than fixed ones when it comes to product life extension policies. Permit decreases in the production of manufactured goods. Provide provisions in distribution agreements for the conditions to be renegotiated in order to take into account an anticipated decline in unit sales. New ties may need to be formed, and subsequently, old relationships may need to end in order to accommodate new procedures and supply partners. Before committing to new providers, businesses should carefully analyze the termination conditions in their existing agreements.

6.3.9 Customer Assertions

Disgruntled customers may file lawsuits for deception or on the basis of their consumer rights if claims made in marketing materials regarding a product's extended lifespan prove to be false. Customers may claim, for instance, that a product intended for planned obsolescence—a design that makes it obsolete quickly after purchase—is unfit for its intended use. In the fast-paced world of product development, this can occur when repairs are not supported for an extended amount of time after a product is out onto the market. Claims are already starting to be made and some of these acts are already being recognized. But regulations are starting to take effect in this sector today (Kottmeyer 2021). Unfounded claims made in advertising regarding a product or service's green credentials may also lead to complaints about "green washing" to the appropriate advertising standards body. In these cases, requests for the removal of advertisements or the publication of corrections may be made, which could result in wasted marketing expenses and damage to the reputation of the product or service's manufacturer.

6.3.10 Terms of the Contract

New consumer terms will need to be drafted and additional contracts may be required when companies are switching from a sale of goods model to one of leasing or providing services. These contracts may be needed to engage third parties to perform services and repairs, retrieve goods to the manufacturer's premises, or exchange and resell the original item.

6.3.11 Consequences for Competition

Companies will need to think about how partnerships and initiatives will affect their regulatory compliance when designing new arrangements to meet the aims of the circular economy, especially when it comes to competition laws. Companies may need to exchange some data and information with other parties, maybe including current or prospective rivals, in order to establish up partnerships, joint ventures, product-sharing agreements, and research and development agreements (Halkos and Aslanidis 2024). Even if the goals are to promote sustainability, agreements between current or future rivals that limit competition or actions that misuse a dominating position may still be subject to competition scrutiny.

6.3.12 *An End-of-Life Design Gone Wrong May Now Damage One's Reputation*

For instance, Lucozade's plastic bottles, which were entirely recyclable but had a label that misled trash sorting machinery into believing otherwise, caused widespread uproar, prompting a swift announcement to alter the bottle's design. It is now essential and necessary to have a design that is compatible with end-of-life processing rather than just a "nice to have." How much longer until virgin material packaging is subject to the same scrutiny (Kottmeyer 2021).

6.4 Conclusion

Implementing the circular economy within an organization brings with it a myriad of risks and challenges. One of the foremost considerations is economic risk, particularly the high initial costs and longer return periods when compared to traditional linear models. Securing funding for circular initiatives might require innovative financing mechanisms such as green bonds or grants, as conventional investors may be hesitant. Environmental risks also arise as the transition may lead to unforeseen ecological impacts, such as increased resource consumption or new forms of pollution, necessitating thorough lifecycle assessments and proactive mitigation strategies. Culturally, shifting mindsets away from linear models can be challenging, requiring organizations to foster a culture of innovation and collaboration throughout the supply chain. Social risks, including job displacement and unequal benefits, must also be managed. Companies must engage their workforce and communities, aligning the circular economy objectives with broader corporate social responsibility (CSR) strategies. Additionally, regulatory hurdles may present obstacles as many existing laws do not account for circular principles. Continuous engagement with policymakers and adherence to evolving regulations are crucial to avoiding compliance issues. Strategically, the transition to a circular model requires governance structures that support long-term sustainability goals. The board must ensure that CE strategies are integrated into the company's vision and core values. Clear communication with stakeholders, including customers, suppliers, and regulators, is essential for maintaining alignment and minimizing resistance. Furthermore, fostering partnerships across industries, including academia and government entities, can accelerate innovation and reduce risks. In conclusion, while the journey to adopting the circular economy is fraught with risks across economic, environmental, social, and regulatory dimensions, it also offers significant opportunities for innovation, sustainability, and long-term value creation. Organizations that successfully manage these risks and align their business strategies with circular principles will be better positioned to thrive in an increasingly resource-constrained and environmentally conscious global market.

References

Agyemang M et al (2019) Drivers and barriers to circular economy implementation: an explorative study in Pakistan's automobile industry. Manag Decis 57(4):971–994

AlMashaqbeh S, Munive-Hernandez JE (2023) Risk analysis under a circular economy context using a systems thinking approach. Sustain For 15(5):4141

Awan U, Sroufe R (2022) Sustainability in the circular economy: insights and dynamics of designing circular business models. Appl Sci 12(3):1521

Blomsma F, Tennant M, Ozaki R (2023) Making sense of circular economy: understanding the progression from idea to action. Bus Strateg Environ 32(3):1059–1084

Choudhary D, Kumar R (2024) Risk investigation in circular economy: a hierarchical decision model approach. Int J Log Res Appl 27(1):103–128

Dan A et al (2023) Risk analysis of adopting the circular economy practices: a perspective of resource-dependent theory. Supply Chain Forum: An International Journal 25:504. Taylor & Francis

dos Santos Leite Gonella J et al (2024) A behavioral perspective on circular economy awareness: the moderating role of social influence and psychological barriers. J Clean Prod 441:141062

Dulia EF et al (2021) Admitting risks towards circular economy practices and strategies: an empirical test from supply chain perspective. J Clean Prod 317:128420

Gennari F, Cassano R (2020) Circular economy and strategic risk. In: Symphonya emerging issues in management. Ist. di Economia d'Impresa, Milano, pp 136–148

Halkos GE, Aslanidis P-SC (2024) How waste crisis altered the common understanding: from Fordism to circular economy and sustainable development. Circ Econ Sustain 4:1–25

Hassan ST et al (2024) The heterogeneous impact of geopolitical risk and environment-related innovations on greenhouse gas emissions: the role of nuclear and renewable energy in the circular economy. Gondwana Res 127:144–155

Kottmeyer B (2021) Digitisation and sustainable development: the opportunities and risks of using digital technologies for the implementation of a circular economy. J Entrep Innov Emerg Econ 7(1):17–23

Obiuto NC et al (2024) Implementing circular economy principles to enhance safety and environmental sustainability in manufacturing. Int J Adv Multidiscip Res Stud 4(2):22–29

Patra SP, Wankhede VA, Agrawal R (2024) Circular economy practices in supply chain finance: a state-of-the-art review. BIJ 31(7):2197–2216

Shooshtarian S et al (2024) Circular economy in action: the application of products with recycled content in construction projects–a multiple case study approach. Smart Sustain Built Environ 13(2):370–394

Singha R et al (2024) Exploring challenges of circular economy. In: Smart cities and circular economy: the future of sustainable urban development. Emerald Publishing, Leeds, p 59

Souza Piao R et al (2024) Barriers toward circular economy transition: exploring different stakeholders' perspectives. Corp Soc Responsib Environ Manag 31(1):153–168

Stylianou M et al (2024) Analyzing the SWOT of circular economy development in established industrial zones: a case study from Cyprus. Sustain Chem Pharm 39:101513

Chapter 7
Circular Supply Chain Business Models to Incorporate Circularity Concept

Abstract The concept of circular supply chain business models integrates sustainability into corporate operations, aiming to minimize environmental impact and optimize resource use. This study explores the circularity concept by examining strategies such as reducing waste, reusing materials, and recycling products, structured under the 9Rs framework: Refuse, Rethink, Reduce, Reuse, Repair, Refurbish, Remanufacture, Repurpose, Recycle, and Recover. These strategies provide a hierarchy for sustainable business practices, though their effectiveness remains debatable due to insufficient empirical support. The study emphasizes the challenges industrial companies face in implementing circular business models, such as aligning operational limitations with available resources and capabilities. The paper presents three circularity-related strategies: keeping ownership of the product (RPO), extending product life (PLE), and designing for recycling (DFR), and discusses their implications for supply chain resilience and economic viability within the framework of the circular economy.

Keywords Circular supply chain · Sustainability · 9Rs framework · Recycling · Product life extension · Resource optimization

7.1 Introduction

A sustainable company plan may be formulated using a circular business model as a framework. The 9Rs: Refuse (make product redundant), Rethink (intensify the product's use), Reduce (increase efficiency in production or use), Re-use, Repair, Refurbish, Remanufacture, Repurpose (new product with different function), Recycle and Recover (incineration of materials with energy recovery) are some examples of how circular business models formulate these sustainable business strategies (Lit et al. 2024). There is an implied hierarchy—a desired arrangement of sustainable business strategies—within these strategies. The 9Rs are a series that represents the hierarchy of sustainable business strategy. According to the hierarchy, company strategies at the top of the hierarchy are likely to be more sustainable.

J. Kandasamy et al., *Circular Supply Chain Management*, Circular Economy and Sustainability, https://doi.org/10.1007/978-3-031-81750-2_7

Nevertheless, there is currently insufficient empirical support for the hypotheses arguing that methods further up the ladder actually result in more sustainable businesses. Conversely, a number of real-world instances suggest that recycling is not as sustainable as burning or disposing of things. This is due to the fact that recycling necessitates both product disassembly and reverse logistics for product collection from end users. Energy, water, chemicals, production, and distribution capacity will all be used as a result. In some circumstances, this will have a greater financial and environmental effect than if virgin resources had been utilised and the product had been burned after its useful life (Rangan et al. 2023). Merely depending on presumptions about the hierarchy of sustainable business strategies is insufficient for evaluating sustainable business strategies. Rather, the ability to gauge the real performance of procedures once a plan has been implemented is required.

A growing number of industrial firms are discussing the concept known as the "circular economy," This enables them to establish supply networks that recycle or recover the resources used to manufacture their products. CEOs are obviously lured to the concept of lowering their environmental footprint, eliminating operational waste, and optimizing the use of expensive resources (Solomon et al. 2024). However, developing a circular business model is difficult, and making a mistake may cost a lot of money. The source contend that a number of elements contribute to success, but that selecting a strategy that takes into account the company's operational limits as well as its capabilities and resources may be the most crucial. It's natural that a rising number of industrial companies are talking about the circular economy, which pushes for logistics systems that reclaim or reuse the resources used in product manufacturing. CEOs are obviously lured to the concept of lowering their environmental footprint, minimizing operational waste, and optimizing the use of expensive resources. Interface worked hard to ensure the success of this historic move, even establishing a network of carpet wholesalers to service clients across the United States on its behalf (Brändström et al. 2024). This effort was intended to shut the loop in the commercial carpeting supply chain. Despite seven years of strong sales efforts, Interface has only attracted a limited number of lessees. The great majority of customers preferred to buy their carpets rather than rent them since carpet maintenance was included in the janitorial services category and hence invisible to them. They struggled to comprehend the benefits of paying relatively high monthly fees. The ESA programme just lacked scalable capabilities.

In 2000, Interface underwent a strategic transformation, shifting from leasing to manufacturing modular carpet tiles with a strong emphasis on sustainability. By incorporating recyclable vinyl backings and nylon fibers, among other eco-friendly elements, the company managed to cut carbon emissions during production by an impressive 75% compared to industry norms (Groenewald 2024). Furthermore, Interface's adoption of renewable energy sources at its production facilities led to an estimated 69% reduction in its overall carbon footprint. This strategic pivot enabled Interface to capitalize on its manufacturing expertise while prioritizing recycling initiatives, thereby solidifying its position as a leader in environmentally conscious flooring solutions.

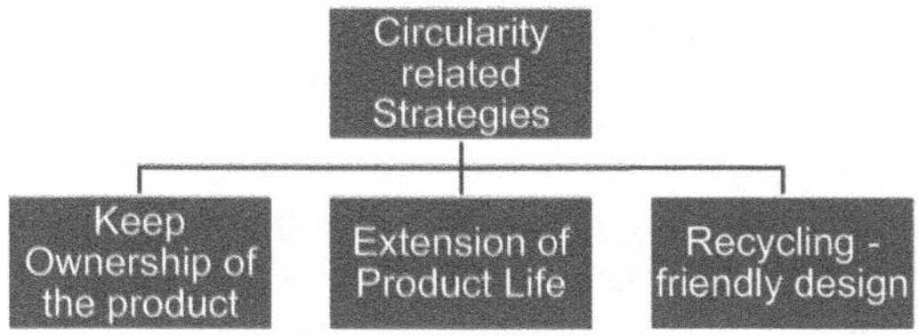

Fig. 7.1 Circularity related strategies

On Interface's content, developing a sustainable circular business model involves a number of considerations, the most crucial of which is deciding on a course of action that fits a company's resources and skills while also taking operational limitations into account (Machado et al. 2024). The three fundamental circularity techniques are outlined in the pages that follow, along with a tool to assist manufacturers in determining which is most likely to be financially viable for them. Our suggestions are based on decades of study and discussions with several global firms.

7.2 Three Circularity-Related Strategies

Manufacturing businesses may develop a circular business model in a variety of ways. These businesses can range from those that make goods for the new economy to those that are more conventional and supply our apparel and furniture (Nishal et al. 2023). Most use a mix of these three fundamental tactics as shown in Fig. 7.1.

7.2.1 Keep Ownership of the Product (RPO)

Instead of selling the goods to the consumer, the manufacturer rents or leases it to them in the traditional version of this strategy. Therefore, once customers have done using a product, the maker is still liable for it. RPO is a compelling approach for businesses that provide intricate goods with a high degree of embedded value (Rangan et al. 2023). Xerox is a prime example; it has long leased printers and photocopiers to business clients. In comparison to a sell and replace approach, this one would need businesses to make significant investments in after-sales and maintenance skills, which could be more costly for both them and their clients in the long run. When simpler items are quite expensive and infrequently needed, RPO can also be used. For example, prom attendees have been renting tuxedos for decades, and this is a trend that is growing in popularity as society grows more cognizant of status. For instance, those in need of a classy ensemble for a one-time function might rent designer clothing from Rent the Runway, an online fashion subscription service. In terms of their basic materials, for instance, its clothing may have minimal intrinsic worth, but its brand value might be quite high (Salehi et al. 2024).

7.2.2 Extension of Product Life (PLE)

Businesses using this tactic concentrate on creating longer-lasting products, which could create opportunities for markets for second-hand goods. For original equipment manufacturers, this might seem like a bad idea because a longer product lifespan eventually translates into fewer purchases. However, as we've seen with the outdoor apparel company Patagonia and the high-end home appliance company Miele, durability is a crucial differentiator in the marketplace and offers a compelling case for premium pricing. Additionally, PLE can assist businesses in stopping clients from switching to other brands (Ada et al. 2024). For instance, Bosch Power Tools can compete with new products from low-cost, low-quality manufacturers by remanufacturing its used tools to increase their lifespan.

7.2.3 Recycling-Friendly Design (DFR)

Businesses that use this approach rethink their goods and production procedures to increase the amount of materials that can be recovered and used to create new goods. This tactic frequently entails forming alliances with businesses that may be most suited to utilise the materials recovered or that possess specialised technological know-how (de Koning et al. 2024). One example is Adidas's six-year collaboration with Parley for the Oceans. Adidas produces their shoes and clothing using textile thread that Parley makes from scrap plastic. Their collaboration lessens the quantity of plastic debris in the world's waters. It's important to ask some very specific and practical questions to find out which combination of the three basic strategies will bring your company the most value (Meyer et al. 2024). Some examples of these questions include whether you can move your product, remanufacture it, or recover it from the customer. Now let's examine the ideal format for that conversation.

7.3 The Matrix of Circularity

A circular business model can only be sustained if the product has economic value. It can be achieved by recycling the product, increasing the value of the raw materials and energy used in its manufacture, or by disassembling it into raw materials or component parts that can be recycled for different purposes. Of course, value doesn't have to be material; as the fashion industry has shown, a circular business model can be developed where value is nearly totally intangible. The potential for developing a circular business model around a product is often correlated with the value inherent in that product, whether it's through its brand reputation, the resources utilized in its production, or the premium customers are willing to pay for sustainability (Papamichael et al. 2024). The extent of value manufacturers can extract from a circular model is influenced by external factors such as legislation, the presence of active commodity component markets, and the availability of secondary markets for used goods. Products with high inherent value that are easily obtainable and manageable represent ideal candidates for circularity, requiring minimal adjustments to the existing business model. Gorenje, a Slovenian white goods company, tested a circular business model by leasing washing machines, leveraging their durable nature and potential for extended use. However, customers found that the costs associated with leasing, including expected services, outweighed Gorenje's maintenance investments, leading to a perception that leasing was more expensive than ownership (Saari et al. 2024). This case illustrates the challenges and consumer perceptions that can impact the success of circular business models, emphasizing the need for careful market adaptation and consideration of cost dynamics.

7.4 Viability of Circularity Approach

Evaluating the viability of a particular circularity approach necessitates a thorough analysis of costs and benefits as well as some trial and error (Schrijer 2024). But businesses can make their thinking clearer by responding to these two questions:

7.4.1 Return the Merchandise

Over 97% of plastic bottles in Norway are recycled. This is due to two factors: first, the remarkably high level of enthusiasm for recycling among Norwegians; second, a vast retail network supporting a government-managed deposit-refund programme exists for bottle collection, including reverse vending machines. Put differently, Norway possesses two essential components for the successful operation of reverse supply chains: infrastructure and public involvement. In the absence of them, it can be difficult to obtain used goods for circularity, as seen by the significantly lower rates of recycling and collection of plastic bottles in other regions of the world. The existence of secondary markets for used goods and commodity markets where extracted raw materials can be sold is another factor to take into account. It will be hard for original sellers to close the circle when consumers are hesitant to part with used goods that have a high resale or exchange value, like power tools and construction equipment. In those situations, obtaining used goods might necessitate costly buyback or trade-in schemes, which is precisely why businesses frequently take leasing as a viable business strategy: If you own products, recovering them is simpler.

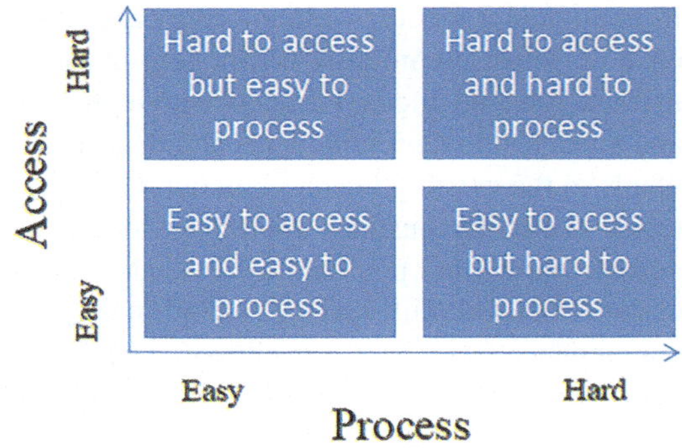

Fig. 7.2 Strategic options for circular business model

7.4.2 *Obtain Product's Value*

Products that are very large or heavy, or that contain potentially dangerous materials, may be simple for manufacturers to legally reclaim, but they may be costly and difficult to relocate and recondition. Moving a washing machine, for instance, will undoubtedly be more difficult and costly in a reverse supply chain than moving an ink cartridge. Value recovery is also a challenge for intricately constructed products. More modular, coarse-grained devices, like desktop computers, are easier to reassemble than sophisticated smartphones and laptops. Lastly, the availability of affordable options for product reformulation will determine whether value recovery is feasible. Should labour-intensive manual work be necessary, the residual value of the used goods must be sufficient to warrant the expenditure.

7.5 Strategic Options for Developing Circular Business Model

In a two-by-two matrix that shows the strategic options for developing a circular business model as shown in Fig. 7.2, the answers to those two questions aid in determining where a business fits in. Now let's examine each quadrant (Stabler et al. 2024).

7.5.1 *Hard to Get to and Difficult to Handle*

Particularly for products with relatively little value locked up in them, products in the top right quadrant may experience a level of wear and tear that makes simple repair and remanufacturing impossible, eliminating the possibility of extending product life. It may be more difficult for the manufacturer to obtain those that are in good shape and still functional in customer-to-customer secondary markets due to their high resale value. We recommend that manufacturers take into account a combination of design for recycling and maintaining product ownership in order to prevent the loss of the value embedded in them. Though it hasn't stated it clearly, the French tyre company Michelin appears to be moving in that direction. The company has pledged to employ 80% sustainable materials in its manufacturing process and has also purchased Lehigh Technologies, a Georgia-based manufacturer of environmentally friendly rubber powders made from recycled tyres. In the meantime, it is making a shift to RPO in many B2B markets, offering cost savings on tyre leasing and performance improvements for commercial fleets (Stank et al. 2019). Modern information technologies are used by Michelin's leasing businesses to deliver those benefits. For instance, its Effifuel division equips cars with digital sensors to track performance indicators like fuel economy and carbon emissions.

Going circular could have little value potential for businesses in this quadrant, even if their pricey products have a lot of valuable ingredients. An extreme example are wind turbines, which require a lot of resources to create but are difficult to access due to their remote locations and size, and their complex architecture and composition make it difficult to extract reusable materials or components. Wind turbines are accumulating in desert landfills for this reason. That is not to say that turbine companies should abandon circularity; rather, they should start by thinking about how extending the product's life can unlock value. Companies may pave the way for an RPO model, in which turbine manufacturers—rather than power companies—retain ownership and responsibility for the turbines, by investing in durability and modularity to make products last longer and be easier to maintain. Longer term, they may be able to implement a DFR strategy, which would reduce the turbines' dependency on non-recyclable materials and facilitate their disassembly.

7.5.2 Easy to Access But Hard to Process

Generally low-embedded-value goods like carpets, mattresses, and sports shoes are located in the bottom right quadrant. On the one hand, there's a good chance that these products can be recovered from the customer due to their portability and lack of a profitable secondary market. However, they are difficult to recondition and require complicated material extraction techniques. In these situations, designing products for recycling will be necessary to go circular. The 2014-founded Dutch start-up DSM-Niaga created carpets made of pure polyester instead of the industry-standard materials, which contain an indissoluble complex of chemicals including numerous known carcinogens, and a fully recyclable mattress made up of six modular components. When a mattress reaches the end of its useful life, its basic materials can be recycled and it is simple to disassemble for easier maintenance (the covers, for instance, are washable and removable) (Shabanpour et al. 2024). Carpets' uniform design greatly speeds up material recovery and lowers the amount of water and energy required during production. Also, the new carpets are lighter, which facilitates handling and transportation.

Companies even very large ones may need to find partners to make circularity work when products have comparatively low value. Adidas is one example, as was previously mentioned. Nike, its main rival, also collaborates with others: Used athletic shoes are recycled by Nike Grind into raw materials for brand-new products like Future Foam's carpet padding and AstroTurf's playing fields. Initiatives of a similar nature are observed in other sectors, such as electronics and consumer goods. This quadrant's products include a lot of small, high-tech gadgets with a lot of highly integrated components, hard-to-extract toxic materials, and short life cycles, like smartphones (Yanamandra 2019). These products usually have active secondary markets, which helps the environment overall but makes it more difficult for the original manufacturers to access them. For the leading producers of sophisticated consumer electronics, that poses less of an issue. They frequently take on the role of the pillars in an ecosystem within the industry, lowering access barriers for manufacturers by erecting obstacles to exit for device owners. Apple is arguably the best example. At Apple's own retail locations, customers purchase, fix, and trade in their devices; as a result, Apple owns the data generated by their use of those devices. Naturally, there is a robust secondary market for iPhones and iPads, which would normally make it more difficult for the company to recover its investment. However, the quick release of next-generation products and trade-in rebates encourage customers to part with their old Apple devices in exchange for cheaper upgrades.

7.5.3 Hard to Obtain But Easy to Process

In the upper left quadrant of goods with harder recovery due to usage, strategies like Design for Recycling (DFR) are crucial. For instance, BioPak, an Australian food-packaging company, implemented entirely biodegradable packaging alongside a composting service spanning multiple postal codes. This approach diverted 660 tons of waste from landfills by enabling customers to dispose of biodegradable packaging with food scraps for collection. As embedded value increases, access to items can become challenging, leading to strategies like Remanufacturing Product Ownership (RPO). Caterpillar exemplifies this with its Cat Reman program and Cat Certified Rebuild service, offering remanufactured parts and equipment to prolong their lifespans. Additionally, Xerox employs a comprehensive circular strategy encompassing Pay-Per-Use, Product Lifetime Extension (PLE), and DFR. Through modular design and aggressive material recycling, Xerox keeps copiers in operation longer while reducing virgin resource inputs. These examples highlight the diverse approaches companies employ to enhance sustainability and circularity across different product categories.

7.5.4 *Easy to Obtain and Easy to Process*

In the lower left quadrant, products with well-established recycling systems, like plastic bottles in Norway or aluminum globally, benefit from strategies that extend and streamline existing processes. Real Alloy in the US and Norsk Hydro in Oslo are examples of companies refining recycling efficiency by minimizing industrial by-products. These incremental improvements enhance recycling systems, making them more efficient and profitable. High-value, easily obtainable items, such as those with strong brand recognition, present ideal opportunities for circularity with minimal business model changes. Patagonia's Worn Wear program, for instance, leverages its eco-conscious image to create a market for used clothing, combining Product Lifetime Extension (PLE) and Design for Recycling (DFR) principles. Ultimately, success in the circular economy hinges on executive commitment, organizational adaptability, and ongoing technological and regulatory developments. The circularity matrix provides a tool for companies to assess their approach based on resources, capabilities, and market conditions, facilitating strategic decision-making in pursuit of sustainability goals.

7.6 How Does Supply Chain Resilience Benefit from the Circular Economy?

In a conventional linear economy, goods are produced, consumed, and then discarded. Despite the traditional manufacture, use, and dispose model, assets in a circular economy are utilized for as long as possible to maximize their worth (de Koning et al. 2024). Once an object approaches the end of its useful life, its components are recovered and recycled.

Three principles underpin the circular economy and enhance supply chain resilience:

1. Simplify, eliminate toxicity, and cut waste from products for better end-of-life resource management.
2. Preserve raw materials at optimal quality and extend their useful life.
3. Reintroduce elements that have a beneficial effect on the environment.

Implementing the circular economy holds immense promise, offering potential economic gains in the trillions of dollars, significant job creation, and substantial emissions reductions. However, forward-thinking supply chain organizations have identified three key challenges. Design stage lock-in of raw material selections: Traditional product design often commits to specific raw materials early in the process, hindering circular practices. Overcoming this hurdle requires innovative design approaches that prioritize recyclability and reusability from the outset. Increasing demand for end-of-life materials management: A shift towards a circular economy necessitates new business models and improved consumer engagement to effectively handle end-of-life materials (Meyer et al. 2024). This entails developing strategies for collection, recycling, and repurposing, as well as educating consumers on proper disposal practices. Lack of stringent selection criteria in collaborative partnerships: While collaborative partnerships are vital for circular economy initiatives, a lack of clear selection criteria can lead to excessive resource investment with minimal returns. Establishing strict criteria ensures that partnerships are productive and mutually beneficial. Addressing these challenges demands innovative solutions, cross-sector collaboration, and supportive policies to facilitate the transition to a circular economy.

7.6.1 First Action Item

Promote the integration of circular economy principles into product design from the outset. Early selection of raw materials in the design phase can significantly impact refurbishment and end-of-life product management processes. To proactively influence decisions regarding raw materials, supply chain organizations should act as collaborative partners during the design phase. This involves contributing expertise to material selection, identifying opportunities for material innovation, and co-authoring design guidelines tailored for the circular economy. By engaging in these activities early on, organizations can pave the way for more sustainable and circular product life cycles.

Common approaches include:

1. One common approach to promoting the circular economy involves breaking down silos within the supply chain to collaborate with design teams. By fostering cross-functional communication and collaboration, organizations can identify opportunities to create value by transitioning towards circular practices. .
2. Another is to take action by putting in place common design principles to satisfy business objectives and the outcomes of the circular economy.
3. Collaborate with design teams to find innovative raw material partners by posing open design challenges. Find chances to work with R&D to invest in innovative materials.

While establishing specific goals for the incorporation of recycled content into new goods to boost the secondary materials market, evaluate stakeholders, paying particular attention to customer acceptance requirements. Establish a progressive commitment over time, keeping in mind any possible detrimental effects on the caliber of the final product and the customer's experience using it.

7.6.2 Second Action

Encouraging client involvement in the circular economy is essential for promoting sustainable practices. Businesses are adopting product-as-a-service models, relying on consumers to return items for recycling or refurbishing. Additionally, third-party standards serve as environmental scorecards, aiding consumers in making informed purchase decisions based on a product's sustainability. By engaging clients and providing transparent information, businesses can drive positive change towards a more circular and environmentally friendly economy. To extend the lifespan of items, companies are increasingly providing maintenance and repair services. Extended product warranties or pay-per-repair models are two ways to do this. Among the actions supply chain organizations can take are:

1. Collaborate with sales to evaluate the potential for portfolio segmentation by utilizing external standards or internally defining the items' sustainability and circularity credentials.
2. Engage with sales to determine the amount of acceptance and evaluate the impact of selling items as a service or reconditioned products for sale on supply chain network design.
3. Identify prospects for strategically raising the rates of raw material recycling by evaluating the end-of-life consumer behavior of goods and collaborating with industry partners.

7.6.3 The Third Action Sets Explicit Standards for Collaborations in the Circular Economy

A circular economy is a system shift that depends on teamwork to solve issues and reach a larger size. According to the research, 77% of supply chain companies work with international organizations, foundations, specialized recyclers, startups, and other entities. In order to establish scalable circular economy solutions, organizations are now starting to collaborate across industries. It takes time and resources to collaborate. Leaders in the supply chain need to be transparent about the resources needed and the anticipated benefits of working together. Among the requirements are

1. Identify the selection and success criteria for cooperative partnerships by matching cooperative efforts with the goals of the supply chain circular economy. As part of the selection criteria for cooperation, take into account the abilities and desire to learn of the collaborator as well as potential for shared value.
2. Utilize partnerships to effectively address obstacles to the circular economy by keeping an eye out for new developments, technologies, and collaborators.
3. Regularly evaluate partnerships by evaluating the benefits they have brought to the advancement of the circular economy plan and the removal of obstacles.

7.7 Models of Supply Chains for the Circular Economy

7.7.1 Models of Products as a Service

The necessity for consumer cooperation in recovering items for recycling or reuse presents a significant challenge in creating a circular supply chain. To address this challenge, some businesses are transitioning to product-as-a-service models. For instance, Signify (formerly Philips Lighting) adopts this approach, granting them control over their products throughout their lifecycle (Shabanpour et al. 2024). By offering products as services, businesses can maintain ownership and responsibility, ensuring better management of resources and promoting a more sustainable circular economy.

7.7.2 Recycling With an Incentive

This product-as-a-service concept is one way that businesses avoid having their products end up in dumps. A different approach is to provide rewards to consumers who recycle or return things to the firm. One such example is Vodafone's phone trade-in program. Any old gadget can be traded in for a trade-in value that can be applied toward a new one from the firm. In the meanwhile, the firm may recycle and salvage used phones that customers trade in, cutting down on waste. Something similar is being done by the apparel store H&M (Schrijer 2024). The firm has stated that it wants to form its transaction "circular," and one method it is achieving this is by gathering worn H&M apparel to be recycled into new items. Worn clothing can now be dropped off in collection bins found at H&M stores. H&M says that over half of the materials it uses to make its clothing are now recycled or sourced responsibly, and by 2030, it wants to have reached 100%.

7.7.3 Pallet Pooling Made of Plastic

Pallet pooling, among the earliest circular business strategies, revolutionized supply chain logistics by enabling companies to rent pallets instead of purchasing them outright. This approach spared businesses the high upfront costs associated with acquiring and managing pallet fleets while reducing waste in the supply chain. Alternatively, they might hire the precise quantity of pallets they require on a daily basis and trust the pooling service to keep the pallets safe and retrieve them at the end of the supply chain. But there were some problems with this concept as well. In the past, pooling firms rented reusable wood block pallets. Although these pallets were more durable than the fragile one- or two-use stringer pallets, they still required regular maintenance and, after around 20 journeys through the supply chain, ended up in the trash. Pallet pooling is made completely circular when plastic pallets are used rather than wooden ones. Plastic is a non-organic material that lasts far longer than wood because it is resistant to mold, mildew, dampness, and insects. Pallets made of plastic are simply crushed down and recycled into new ones when their useful life come to an end. The pallet pooling provider can guarantee that the pallets adhere to a closed-loop model of reuse, repair, and recycling since they have control over each pallet for the duration of its life (Lit et al. 2024).

7.7.4 Using Sustainable Vendors

Logistics outsourcing is a widespread practice in supply chain management nowadays. Before goods and shipments reach their final consumers, they may pass through the hands of several different businesses. This is a problem for businesses wishing to concentrate on a circular business model as they can select suppliers that don't adhere to the same sustainability and waste minimization guidelines. Due diligence is necessary when signing new contracts with vendors and extending current ones for businesses hoping to integrate into a circular economic model. Supply chain managers should incorporate inquiries on ethical, circular, and sustainable business practices into their routine vendor screening procedure. They have to think about looking for businesses that have earned green business certifications (Machado et al. 2024). Green supply chains and green practices complement each other by emphasizing waste reduction and more effective use of natural resources.

Plastic pallet pooling is more than simply a real-world illustration of a circular supply chain. Selecting a provider for plastic pallet pooling entails working with a supplier that adheres to a sustainable business strategy. They may assist your company in achieving its first step toward circularity while serving as an example of reuse and recycling. Pallet pooling demonstrates how circular business practices can benefit companies, customers, and the environment in the real world. Companies can achieve sustainable growth by transitioning from a linear industrial model to a circular supply chain approach. In a circular supply chain, resources are kept in use for longer and may regenerate into new forms at the end of their lifespan. This shift reduces waste and optimizes resource usage, fostering environmental sustainability while driving business growth. Embracing the reverse logistics process can lead to the following advantages.

7.7.5 Reduce Costs and Increase Company Value

Businesses may maximize the value of the raw materials they buy by concentrating on reusable materials. When a product reaches the end of its life cycle, it can be remanufactured and sold again for a profit at a cheaper cost than buying fresh resources to create something new. In this manner, businesses may successfully link the beginning and conclusion of the linear supply chain while lowering total manufacturing costs.

7.8 Advantages of the Circular Economy for Society (Saari et al. 2024)

Almost every industry is still talking about sustainability. It is the responsibility of businesses as well as individual customers to select items with minimal environmental impact as consumer waste grows annually. Customers are becoming more aware of the potential environmental effects of their purchasing selections. 30,000 people in 60 countries were surveyed by Nielsen in 2015, and the results showed that 66% of consumers worldwide said they would be prepared to pay more for sustainable items. If businesses want to maintain their relevance with customers and increase their market share, they should be catering to a public that is becoming more aware of their environmental effect.

7.8.1 Reusing and Recycling Helps Prevent Price Volatility

Because the price of raw materials is continually fluctuating, businesses find it difficult to manage their budgets and keep the overall cost of items under control. Numerous types of virgin materials have daily price fluctuations. For instance, current metal prices have fluctuated more than they have in any twentieth-century decade. Businesses may more precisely estimate their expenses and manage costs by projecting the quantity of recycled and repurposed materials that can be employed in the creation of new products.

7.8.2 Fulfilling Requirements of Regulations

Government policies and regulations pertaining to trash disposal and recycling are driving businesses to adopt more environmentally friendly practices. Some are providing incentives to businesses that actively work to "go green." Reducing the environmental effect is the ultimate aim, regardless of how they choose to increase their profits. In the end, implementing a circular supply chain plan shouldn't be seen as a trade-off between environmental stewardship and sound commercial practices. It's both, businesses that adopt a circular supply chain model stand to earn not just more value from their end-of-life products but also the reputation and intangible advantages of being socially conscious.

7.9 Circular Economy Examples (Brändström et al. 2024; Meyer et al. 2024; Salehi et al. 2024)

7.9.1 *Recycling of Cell Phones*

In addition to cell phone exchange programs, where consumers can earn credit toward a new handset by surrendering an old one, several manufacturers in the market offer clients reconditioned phones at discounted costs. This promotes and encourages the direct return of resources to the makers, strengthening consumer perception and brand loyalty. Businesses can then use these recycled resources to repair outdated technology or recycle its pieces for use in new products.

7.9.2 *Eco-Friendly Style*

Underused clothes or a lack of textile recycling results in value losses of over $500 billion. A movement known as "sustainable fashion" emerged in reaction to the unethical and environmentally damaging practices of fast-fashion retailers. The low cost of fast-fashion apparel obscures the high negative externalities, which are estimated to be worth $192 billion a year. Sustainable fashion employs creative business models that rethink ownership, as well as redesign and repair, to extend the life-cycles of clothes, both for consumers and for the product's quality and reusability. Longer wear cycles are encouraged by sustainable fashion, which also uses materials made from leftover fabric or garments.

7.9.3 *Cross-Sector Collaborations*

When garbage from one business can be used as raw material by another, the circular economy takes off! The Plant in Chicago is one instance with which I am personally familiar. The Plant is a manufacturing and research center that is home to several companies, including a local grocery store, brewery, and chocolate shop. These businesses utilize each other's leftovers to run a zero waste operation. The Plant aims to provide repeatable models of energy and waste loop closures in order to demonstrate the profitability of zero waste enterprises and to persuade others to adopt similar strategies.

7.9.4 Solutions and Business Models for the Circular Economy

In order to effectively implement a circular economy model, your company has to reevaluate its existing systems. Determine the location and timing of waste as well as the best ways to halt any "leakage." Redesigning procedures to guarantee a closed cycle of production puts your company in a successful, sustainable long-term commercial position.

7.10 Conclusion

The transition to a circular supply chain model is a complex but essential endeavour for industries aiming to enhance sustainability and reduce their environmental footprint. This study highlights how circular business models, particularly the 9Rs framework, provide structured approaches for businesses to integrate circularity into their supply chains. However, the hierarchy of these strategies requires further empirical validation, as assumptions about their sustainability potential may not always align with practical outcomes. For instance, while recycling appears environmentally sound, its execution may have higher financial and environmental costs due to factors like reverse logistics and resource consumption during disassembly. Companies like Interface and Gorenje provide case studies that showcase both the successes and challenges of adopting circular models. Interface's shift to modular carpet tiles with recyclable components led to significant reductions in carbon emissions, illustrating the potential of circular business models to drive sustainability. Gorenje's experience with leasing washing machines, however, underscores the importance of market adaptation and consumer perception in ensuring the viability of such models. Three key strategies—keeping product ownership (RPO), product life extension (PLE), and design for recycling (DFR)—emerge as fundamental approaches for implementing circularity. These strategies emphasize the importance of aligning business operations with sustainability goals, taking into account factors such as the ease of product recovery, the complexity of disassembly, and the availability of secondary markets. The study concludes that while there is no one-size-fits-all approach to circular supply chains, businesses must critically assess their operational capabilities, resource constraints, and market conditions to develop successful circular models. In doing so, companies not only contribute to environmental sustainability but also unlock economic benefits by optimizing resource use and reducing waste. Furthermore, the adoption of circular supply chain models offers significant long-term benefits beyond environmental sustainability. By extending product life cycles and enhancing resource recovery, companies can reduce dependency on virgin materials, stabilize production costs, and mitigate risks associated with volatile raw material prices. Additionally, circular business practices promote innovation by encouraging new product designs that are easier to recycle or refurbish. These efforts can also strengthen brand reputation, as consumers increasingly prefer companies with sustainable operations. Overall, the integration of circularity into supply chains fosters resilience, allowing businesses to remain competitive in a market that increasingly values environmental and economic sustainability.

References

Ada E et al (2024) A circular business cluster model for sustainable operations management. Int J Log Res Appl 27(4):493–511

Brändström J, Jazairy A, Lindgreen ER (2024) Barriers to adopting circular business models: a cross-sectoral analysis. Bus Strateg Environ 33:4331–4350

de Koning I, Kassahun A, Tekinerdogan B (2024) Benchmarking circularity in supply chains: A systematic literature review. J Environ Manag 366:121676

Groenewald ES (2024) Circular economy strategies in supply chain management: towards zero waste. Power Syst Technol 48(1):464–480

Lit FC et al (2024) Business model innovation in circular start-ups: Overcoming barriers in the circular plastics economy. Int Small Bus J 42(4):506–550

Machado N, Morioka SN, Gohr CF (2024) Can product modularity boost strategies for circular business models? A visual tool proposal representing benefits and barriers considering expert opinion. J Clean Prod 448:141522

Meyer D, Schaupensteiner N, Riquel J (2024) Business models and ecosystems in the circular economy using the example of battery second use storage systems. Sustain For 16(5):1906

Nishal M, Ram Prasad K, Kumanan R (2023) Digital supply chain paradigm. In: Industry 4.0 Technologies: sustainable manufacturing supply chains: volume 1—theory, challenges, and opportunity. Springer Nature, Singapore, pp 1–23

Papamichael I et al (2024) Thriving in circularity: vitality of business models development in circular (bio) economy. Curr Opin Green Sustain Chem 48:100934

Rangan RP et al (2023) Performance metrics in digital supply chain paradigm. In: Industry 4.0 Technologies: sustainable manufacturing supply chains: volume 1—theory, challenges, and opportunity. Springer Nature, Singapore, pp 159–173

Saari L et al (2024) Circular economy matrix guiding manufacturing industry companies towards circularity—a multiple case study perspective. Circ Econ Sustain 4:1–26

Salehi N et al (2024) Towards circular manufacturing systems implementation: An integrated analysis framework for circular supply chains. Sustain Prod Consum 51:169–198

Schrijer R (2024) A consumer and supply chain analysis on the circularity of heat pumps: investigating the potential implementation of a circular business model. MS thesis

Shabanpour H et al (2024) A breakthrough in circular economy: using a closed-loop framework to assess the circularity of supply chains. J Clean Prod 444:141265

Solomon NO et al (2024) Circular economy principles and their integration into global supply chain strategies. Financ Account Res J 6(5):747–762

Stabler D et al (2024) Aligning servitization and circularity: The role of institutional confluence in sustainable business models. J Clean Prod 462:142666

Stank T et al (2019) Toward a digitally dominant paradigm for twenty-first century supply chain scholarship. Int J Phys Distrib Logist Manag 49(10):956–971

Yanamandra R (2019) A framework of supply chain strategies to achieve competitive advantage in digital era. In: 2019 international conference on digitization (ICD). IEEE

Chapter 8
Decision Making Models in Circular Supply Chain

Abstract Decision-making is a crucial process within the circular supply chain, where sustainability and resource efficiency are prioritized throughout a product's life cycle. This study explores decision-making models in circular supply chain management (CSCM), emphasizing how digital technologies and life cycle assessments (LCA) can optimize the process. The integration of these tools enables informed decisions, helping organizations to reduce waste, manage resources, and promote closed-loop systems. Critical factors such as supplier selection, product design, reverse logistics, and inventory management are examined to show how they influence sustainable operations. The paper also highlights the importance of collaboration between stakeholders, the use of digital technologies like IoT and blockchain, and compliance with regulatory requirements to support circular practices. These decision-making models aim to enhance the overall performance of the circular supply chain by promoting recycling, reusability, and reducing environmental impact.

Keywords Circular supply chain · Decision making · Life cycle assessment · Digital technology · Industry 4.0 · Sustainability

8.1 Introduction

The process of gathering information, assessing solutions, choosing a decision from the available choices is known as decision making. Decision-making in the supply chain within a circular economy framework involves making choices that prioritize sustainability, resource efficiency, and the reduction of environmental impact throughout the entire product lifecycle (Orji and Ojadi 2023). This requires a holistic and strategic approach to supply chain management. Organizations need to align their strategies with sustainability goals, regulatory requirements, and the principles of the circular economy for the steps in decision making process as shown in Fig. 8.1. Therefore, the adoption of circular practices in SCM is necessary for reducing the activities that affect the business negatively on the surroundings (Khan et al.

J. Kandasamy et al., *Circular Supply Chain Management*, Circular Economy and Sustainability, https://doi.org/10.1007/978-3-031-81750-2_8

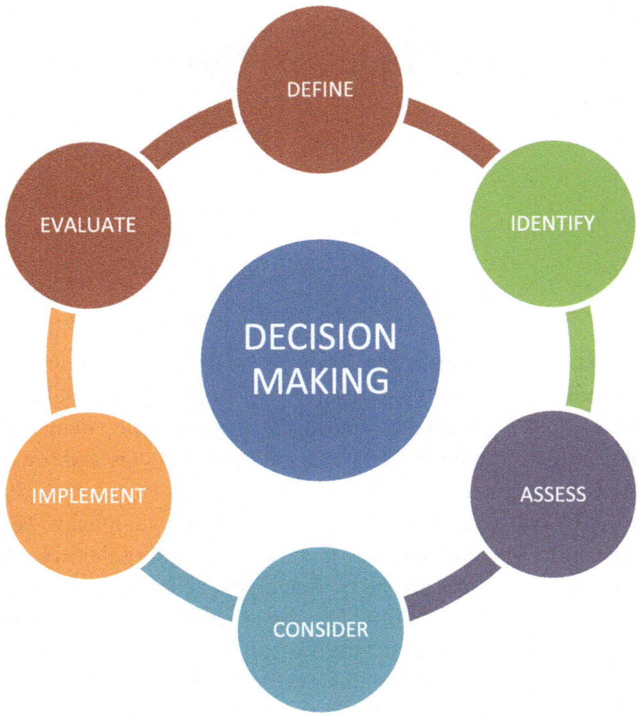

Fig. 8.1 Steps in decision making process

2021; Ncube et al. 2022). The integration of these technologies empowers organizations to make informed decisions based on real-time data, predictive analytics, and simulations. This not only enhances operational efficiency but also supports the implementation of sustainable and circular practices within the supply chain.

8.2 Role of Decision-Making Circular Supply Chain

Circular supply chain is a system of production, distribution, and consumption that aims to reduce waste, decrease environmental impact, and increase resource efficiency (Dwivedi et al. 2023). In the context of supply chain management and circularity, decision-making plays a crucial role in ensuring sustainability, resource efficiency, and the effective implementation of circular economy principles. The circular economy aims to minimize waste and make the most of resources by promoting product reuse, recycling, and the reduction of environmental impact. According to (Pattnaik et al. 2021), the research body addressing the important factors that impact the implementation of a supply chain remains scares. Here are some key aspects of decision-making in supply chain circularity:

8.2.1 Designing Circular Products and Processes

Decision-makers need to consider the entire lifecycle of a product when designing it, ensuring that materials are easily recyclable or reusable. Choices regarding product design, manufacturing processes, and materials impact the feasibility of circularity (Bals et al. 2022) A product's impact on environment should not be considered in its initial phase alone but also after its end-of-life phases such as resold or repaired to extend its working life.

8.2.2 Supplier Selection and Collaboration

Decisions about selecting suppliers play a role in ensuring a circular supply chain. Suppliers who prioritize sustainable practices and materials need to be identified and chosen as they contribute to circularity. Collaboration with suppliers on recycling programs, waste reduction, and closed-loop systems is also crucial. Collaboration with suppliers will help to reduce the recycling or refurbishing cost. The seven steps involved in selecting a supplier are

(i) Identify the needs of the organization
(ii) Constitute criteria for evaluation
(iii) Screen the suppliers
(iv) Request for proposals
(v) Evaluation and selection of suppliers
(vi) Inspect supplier sites
(vii) Finalize the agreement and relationship with the supplier

8.2.3 Logistics and Transportation

Efficient transportation and logistics contribute to reducing the environmental impact of the supply chain. Decision-makers need to choose transportation modes and routes that minimize emissions and resource consumption. The optimization of distribution networks and the consolidation of shipments can contribute to circular supply chain goals.

8.2.4 Inventory Management

Effective inventory management reduces waste and ensures that products are used or sold before they become obsolete or unsellable. Decision-makers need to implement strategies such as just-in-time inventory, demand forecasting, and inventory optimization to minimize excess stock and waste.

8.2.5 Reverse Logistics

Decision-making in reverse logistics is critical for handling product returns, recycling, and refurbishment. Establishing processes for the return, refurbishment, and recycling of products can contribute to the circularity of the supply chain. A Multi-Objective Mixed Integer Linear Programming Model (MO-MILP) can be used to create a reverse logistics model (Kilic et al. 2023).

8.2.6 Technology and Innovation

Decision-makers need to leverage technology and innovation to implement circular economy practices. This may include the use of blockchain for traceability, IoT devices for real-time monitoring, and data analytics for informed decision-making. Investing in technologies that support circularity, such as remanufacturing equipment or recycling facilities, is important.

8.2.7 Regulatory Compliance

Decision-makers must stay informed about and comply with regulations related to environmental sustainability, waste management, and circular economy practices Catherine (Weetman 2016). Compliance with standards and regulations ensures that the supply chain operations align with broader sustainability goals.

8.2.8 Risk Management

It is common in various industries to use risk assessment as an important factor in decision making. During uncertain times, risk management plays a key role. Business activity has always been surrounded by assessing risk. it is necessary to put in place a risk management plan (Settembre-Blundo et al. 2021) Managing risks is important to circular supply chain. Decision-makers need to consider potential challenges related to material availability, market demand for recycled products, and regulatory changes.

8.3 Decision-Making Factors in Circular Supply Chain

The decision-making factors in a circular supply chain are diverse and cover various aspects of the supply chain lifecycle (Farooque et al. 2019). These factors are crucial for organizations aiming to adopt circular economy principles and create sustainable, closed-loop systems. Here are key decision-making factors in a circular supply chain:

8.3.1 Product Design

Materials that are recyclable, reusable, or biodegradable must be preferred and chosen for circular supply chain. These materials may be costlier than the other materials without recycling properties. Components must be designed with modularity to facilitate easy disassembly and repair.

8.3.2 Supplier Selection

Supplier selection is another key important factor in implementing the circular supply chain. The important step in supplier selection process is the purchasing process which is the basis for many other activities (Taherdoost and Brard 2019). A supplier must be assessed and selected based on their commitment to environmental sustainability and circular economy principles. The purchasing manager should find an effective process for identifying the qualified suppliers to suit the business. Suppliers who provide transparent information about the origin and sustainability of materials must be chosen.

8.3.3 Production Processes

Production processes that minimize waste and energy consumption must be opted to improve resource efficiency. Incorporating processes that enable the remanufacturing of products will extend their lifecycle (Weetman 2016). By optimising the production processes, the circular supply chain efficiency gets increased.

8.3.4 Logistics and Transportation

Optimized transportation routes minimize the emissions and usage of resources. Unoptimized routes will increase the transportation and recycling costs. Cost of the logistics can be reduced by selecting sustainable packaging materials and optimized packaging designs.

8.3.5 Inventory Management

Managing a company's sourcing, storage and sales is known as inventory management. Excess inventory space and waste can be reduced by using accurate demand forecasting. Forecasting helps to plan for the end-of-life of products, including recycling or refurbishment. Few examples of inventory management are

(i) Raw materials
(ii) Parts and components
(iii) End products
(iv) Maintenance and repair supplies
(v) Work-in-progress

8.3.6 Reverse Logistics

Establishing efficient processes for handling product returns, repairs, and recycling will simplify the Circular supply chain process. For reverse logistics to function effectively a network for recycling and managing the reverse flow of materials must be developed.

8.3.7 Technology Integration

Circular supply chain management can function efficaciously by integrating it with current technologies. Leveraging data analytics for informed decision-making in supply chain operations. Incorporating technologies such as IoT, blockchain, and AI can enhance visibility and traceability of the circular supply chain.

8.3.8 *Regulatory Compliance*

Environmental Regulations: Staying compliant with regulations related to waste management, environmental protection, and circular economy practices will help to implement reverse logistics at their respective location. Obtaining certifications that validate adherence to circular economy standards is key in adhering to regulations.

8.3.9 *Collaboration and Partnerships*

Supply Chain Collaboration is required by collaborating with suppliers, customers, and other stakeholders to create a closed-loop system. Partner with organizations to share knowledge, resources, and technologies that support circularity.

8.3.10 *Financial Considerations*

Resources and investments should be allocated towards circular supply chain initiatives. Cost-Benefit Analysis must be carried to evaluate the economic viability of circular practices.

8.3.11 *Risk Management*

Risks must be identified and mitigated at the earlier stages of decision-making process associated with circular supply chain practices. Contingency plans for potential disruptions in material supply, recycling infrastructure, or market demand for recycled products should be developed. Successful implementation of circular supply chain practices requires a holistic and integrated approach, considering these decision-making factors across the entire supply chain ecosystem.

8.4 Life Cycle Assessment (LCA) in Decision Making

Life-cycle assessment (LCA) is a process of evaluating the effects that a product has on the environment over the entire period of its life thereby increasing resource-use efficiency and decreasing liabilities (Lehmacher 2017). It can be used to study the environmental impact of either a product or the function the product is designed to perform. LCA can be a useful tool to guide governments, industries, and end users to reduce raw resource consumption and environmental impacts (Subal et al. 2023).

There is a wide gap between the various methods for the analysis of environmental, social, and economic impact (Dwivedi et al. 2023). LCA is commonly referred to as a "cradle-to-grave" analysis. LCA's key elements are:

1. identify and quantify the environmental loads involved; e.g. the energy and raw materials consumed, the emissions and wastes generated;
2. evaluate the potential environmental impacts of these loads; and
3. assess the options available for reducing these environmental impacts.

Life Cycle Assessment (LCA) is a systematic and comprehensive approach for evaluating the environmental impacts associated with a product, process, or service throughout its entire life cycle (Charter 2018). LCA considers all stages, from raw material extraction and production to use, end-of-life, and disposal. Here's how LCA helps in evaluating environmental impacts:

8.4.1 Systematic Approach

LCA takes a cradle-to-grave perspective, considering all stages of a product's life cycle. It ensures a holistic and systematic evaluation, preventing the omission of significant environmental impacts at different stages.

8.4.2 Impact Categories

LCA assesses environmental impacts across various categories, including energy use, greenhouse gas emissions, water consumption, air pollution, land use, and resource depletion. Different impact categories allow for a comprehensive understanding of the overall environmental profile.

8.4.3 Data Collection and Inventory Analysis

LCA involves collecting data on inputs and outputs at each life cycle stage, creating an inventory of the environmental burdens associated with the system (Lehmacher 2017). The data collected includes raw material extraction, energy use, emissions, waste generation, and other relevant information.

8.4.4 Normalization and Weighting

LCA often includes normalization and weighting steps to provide a relative importance of different environmental impacts (De Angelis et al. 2018). Normalization allows for comparing impacts to regional or global averages, while weighting assigns importance factors to different impact categories based on societal values.

8.4.5 Impact Assessment

The collected data is then used in the impact assessment phase to quantify the potential environmental impacts. Mathematical models and impact assessment methods translate the inventory data into impact scores for each impact category.

8.4.6 Sensitivity Analysis

LCA allows for sensitivity analysis, which assesses the influence of uncertainties and variations in the data on the overall results. Decision-makers can understand how changes in assumptions or data inputs affect the reliability of the results.

8.4.7 Comparative Analysis

LCA facilitates the comparison of different products, processes, or scenarios. Decision-makers can use LCA results to identify alternatives with lower environmental impacts and make informed choices based on sustainability criteria.

8.4.8 Identification of Hotspots

LCA helps identify environmental hotspots, pinpointing stages in the life cycle with the highest environmental impact. Decision-makers can focus their efforts on improving or optimizing these critical areas to achieve more significant overall environmental benefits.

8.4.9 Scenario Analysis

LCA allows for the evaluation of different scenarios or interventions to understand their potential impact on environmental performance. Decision-makers can explore various strategies and assess their effectiveness in reducing environmental burdens.

8.4.10 Communication of Results

LCA results can be communicated to stakeholders in a transparent and understandable manner. The information obtained from LCA can be used for eco-labelling, environmental product declarations (EPDs), and sustainability reporting. By providing a comprehensive and quantitative assessment of environmental impacts, Life Cycle Assessment empowers decision-makers to make informed choices, optimize processes, and contribute to sustainability goals and circular economy principles. It is a valuable tool for businesses, policymakers, and researchers seeking to understand and minimize the environmental footprint of products and processes.

8.4.11 Examples of Successful Implementation of LCA in Circular Supply Chain Decision-Making

Several companies and industries have successfully implemented Life Cycle Assessment (LCA) in circular supply chain decision-making to improve sustainability and minimize environmental impact. Here are a few examples:

1. **Interface Inc. (Carpet Manufacturing):**
   Interface, a global carpet manufacturer, adopted LCA to assess the environmental impact of its products and transition towards a more sustainable and circular business model. LCA helped Interface identify areas for improvement in raw material sourcing, manufacturing processes, and end-of-life considerations. The company has made strides in recycling materials, reducing waste, and designing products with a focus on circularity.
2. **Philips (Consumer Electronics):**
   A multinational electronics company, applied LCA to evaluate the environmental impact of its lighting products. LCA insights led Philips to design more energy-efficient products and optimize packaging to reduce waste. The company also implemented product take-back programs to facilitate the recycling of electronic components, contributing to a circular approach in the electronics industry.
3. **Toyota (Automotive):**
   Toyota has integrated LCA into its decision-making processes to assess the environmental impacts of its vehicles. LCA has influenced Toyota's product

design and manufacturing processes, leading to innovations such as hybrid and electric vehicles. The company has also focused on recycling and reuse of materials in car manufacturing, aligning with circular economy principles.

4. **Unilever (Consumer Goods):**
   A global consumer goods company, has applied LCA to evaluate the environmental footprint of its products. LCA has informed Unilever's decisions in product formulation, packaging, and supply chain management. The company has worked towards reducing water consumption, sourcing sustainable raw materials, and implementing recycling initiatives.
5. **IKEA (Furniture Retail):**
   IKEA has incorporated LCA into its sustainability strategy to assess and improve the environmental performance of its furniture and home products. LCA findings have influenced IKEA's material choices, leading to the use of sustainable and recyclable materials. The company has also implemented circular initiatives, such as furniture take-back programs and designing products with disassembly in mind.
6. **Nike (Apparel and Footwear):**
   Nike has utilized LCA to assess the environmental impact of its footwear and apparel products. LCA insights have guided Nike in making sustainable material choices, optimizing manufacturing processes, and developing products with a focus on recyclability. The company has also embraced circular initiatives, including recycling old athletic shoes to create new products.
7. **Nestlé (Food and Beverage):**
   Nestlé has employed LCA to analyze the environmental impacts of its food and beverage products. LCA findings have influenced Nestlé's decisions in sourcing raw materials sustainably, improving production efficiency, and optimizing packaging. The company has committed to achieving zero net emissions by 2050 and enhancing circularity in its packaging materials.

These examples highlight how LCA has been instrumental in guiding decisions related to product design, material sourcing, manufacturing processes, and end-of-life considerations. Successful implementation of LCA in circular supply chain decision-making requires a commitment to sustainability, collaboration across the value chain, and a willingness to adopt circular economy principles.

8.5 Circular Design Thinking

Circular design is the practice of creating durable, reusable, repairable and recyclable products that generate zero waste to support a circular economy (Ramakrishna and Wahab 2023). The circular design aims to accomplish sustainable development by using limited resources and surrounding impacts from economic growth and well-being (Pruhs et al. 2023). The circular design thinking is a type of freedom for companies to make new rules and production methods that meet the environmental directives, (Reslan et al. 2022). In circular design, designers consider all stages of a product's lifecycle and ensure that it is sustainable from manufacturing to disposal. Some examples of the decisions involved in circular design are:

(i) **Raw material:** what is the source material, and how can we extract it sustainably?
(ii) **Product specifications:** is the product reusable, repairable and recyclable?
(iii) **Manufacturing process:** is it energy-efficient, and does it generate any waste?
(iv) **Distribution and use:** How much energy does it take to deliver and use the product/service? Where does this energy come from?
(v) **Disposal:** Can the product be dismantled? Can we use the parts as-is in the manufacturing process again? Can we return the raw materials to the earth?

Integrating circular design principles into decision-making involves incorporating strategies that prioritize sustainability, resource efficiency, and the reduction of environmental impact (Liu and Ramakrishna 2021). Here are steps and considerations for incorporating circular design principles into decision-making processes:

8.5.1 Understand Circular Design Principles

Gain a thorough understanding of circular design principles, which include designing for durability, reparability, recyclability, and reusability. Familiarize yourself with concepts such as cradle-to-cradle design, product life extension, and closed-loop systems.

8.5.2 Life Cycle Thinking

Adopt a life cycle thinking approach, considering the entire life cycle of a product from raw material extraction to end-of-life (Schröder 2019). Analyze the environmental impact at each stage and identify opportunities for improvement.

8.5.3 Collaboration and Stakeholder Engagement

Involve various stakeholders, including product designers, engineers, suppliers, and customers, in decision-making processes. Collaborate with suppliers and partners to ensure alignment with circular design principles throughout the supply chain.

8.5.4 Prioritize Materials Sustainability

Choose materials with lower environmental impact, considering factors such as recyclability, renewability, and biodegradability. Prioritize materials that can be easily disassembled and reused or recycled at the end of the product life.

8.5.5 Modular Design

Embrace modular design principles, allowing components to be easily assembled and disassembled. Modular design facilitates repair and upgrades, extending the product's lifespan.

8.5.6 Design for Disassembly

Ensure that products are designed for easy disassembly, facilitating the separation of different materials during recycling. Use fasteners and connections that are easy to remove without causing damage.

8.5.7 Implement Product Life Extension

Design products with a focus on durability and longevity to extend their life cycle. Consider repairability and provide repair services or guidelines to encourage maintenance and repairs.

8.5.8 Promote Reuse and Refurbishment

Explore opportunities for designing products that can be easily refurbished and resold. Establish programs for collecting and refurbishing used products to promote a circular market.

8.5.9 Closed-Loop Systems

Consider the development of closed-loop systems, where products or materials are returned, refurbished, and reintroduced into the market. Explore partnerships with suppliers and waste management companies to close the loop on specific materials.

8.5.10 Incorporate Digital Technologies

Leverage digital technologies such as Internet of Things (IoT), sensors, and data analytics to enhance product traceability and monitoring. Implement smart design solutions that enable better product lifecycle management and optimization.

8.5.11 Life Cycle Assessments (LCAs)

Integrate Life Cycle Assessments into the decision-making process to quantitatively assess the environmental impacts of design choices. Use LCA results to inform decisions and identify areas for improvement in product design and manufacturing processes.

8.5.12 Continuous Improvement

Establishing a culture of continuous improvement by regularly reviewing and updating design principles based on evolving technologies, materials, and circular economy best practices will improve the process continuously. A feedback loop for continuous learning and refinement of design strategies will help too.

8.5.13 Educate and Raise Awareness

Educate the design team and other stakeholders about the importance of circular design principles. Foster a culture of awareness and responsibility toward sustainability within the organization. By integrating these principles into decision-making processes, organizations can contribute to the development of more sustainable products and systems, aligning with the goals of the circular economy.

8.6 Collaborative Decision-Making in Supply Chains

In a supply chain crisis, you need to involve all stakeholders in decision-making. This means bringing together internal teams, external partners, and customers to collaborate and share information (Srivastava and Sowmya 2023). By doing this, you can make more informed decisions and ensure that everyone is aligned and working towards the same goals.

8.6.1 The Role of Collaboration in Circular Supply Chain Decision-Making

Collaboration plays a pivotal role in the successful implementation of circular supply chain decision-making. Circular supply chains involve a system where materials, products, and waste are managed in a closed loop, emphasizing sustainability, resource efficiency, and the reduction of environmental impact. Here are key aspects highlighting the importance of collaboration in circular supply chain decision-making:

1. **Supplier Collaboration:**
   Collaboration must be made with suppliers to source sustainable and recyclable materials, fostering a closed-loop approach. To have a healthy relationship with the suppliers, transparency is required. Transparency must be built in the supply chain, to ensure that suppliers adhere to environmental and ethical standards.
2. **Design Collaboration:**
   A cross-functional team that includes designers, engineers, and supply chain professionals will integrate circular design principles in an effective way. Products must be designed with modular components, enabling easy disassembly and repair.
3. **Reverse Logistics Collaboration:**
   Collaborating with logistics partners will optimize returns management, facilitating the recycling, refurbishment, or remanufacturing of products. With this collaboration efficient take-back programs for end-of-life products can be conducted for recycling and reusing.
4. **Technology Collaboration:**
   Technology collaboration is the sharing of data with partners and to implement digital solutions for enhanced traceability and visibility across the supply chain. Internet of Things (IoT) devices and sensors can be used for real-time monitoring of products and materials, enabling better decision-making.
5. **Cross-Industry Collaboration:**
   Collaboration with organizations from different industries to share best practices, knowledge, and innovative solutions for circular supply chain management

is known as cross-industry collaboration. Developing partnerships within industry ecosystems will create closed-loop systems that extend beyond individual companies. The more the collaboration, the better.

6. **Collaboration with Regulatory Bodies:**
   Regulatory bodies are government authorities that are responsible for exercising autonomous dominion over few regions of human activity with licensing and regulating capacity. By Collaborating with regulatory bodies, it ensures compliance with environmental standards and regulations related to circular economy practices. Work must be done collectively to advocate for policies that support circular supply chain initiatives.
7. **Customer Collaboration:**
   Customers are the heart of any business. Collaborating with customers will increase their awareness and about the importance of circular products, proper disposal or recycling methods. Feedbacks will help to gather customer insights on product design and circular supply chain improvements.
8. **Collaborative Innovation:**
   Collaboration with research institutions and other companies on joint R&D projects will enhance circular supply chain. Participating in open innovation platforms will give new ideas where companies share challenges and collectively work towards solutions.

Collaboration across the supply chain ecosystem is essential for overcoming the complex challenges associated with circular supply chain decision-making. It enables a more holistic and integrated approach, where stakeholders collectively contribute to the development and implementation of sustainable and circular practices.

8.7 Integration of Technology for Informed Decision Making

The integration of technology is crucial for informed decision-making in various aspects of business, including supply chain management and circular economy initiatives. Here are ways in which technology can be leveraged for informed decision-making in these contexts:

1. **Data Analytics and Business Intelligence:**
   Advanced data analytics tools can be utilised to process and analyze large datasets related to supply chain operations. To do this a business intelligence system is required. Implementing business intelligence systems can generate meaningful insights from data, enabling informed decision-making based on trends and patterns.
2. **Internet of Things (IoT):**
   Deploying IoT devices to collect real-time data from various points in the supply chain, such as warehouses, manufacturing facilities, and transportation will help to choose the right decision. Decision making can be made easy by

monitoring and tracking of products, providing visibility into the entire supply chain for better decision-making.

3. **Blockchain Technology:**
   Blockchain is a process of recording information which is difficult to be changed, hacked, or manipulated. Blockchain can enhance traceability and transparency in the supply chain. Using blockchain to create a secure and immutable record of transactions, ensuring the authenticity of information related to product provenance and sustainability practices.
4. **Predictive Analytics:**
   The process of utilising data to forecast future outcomes is known as predictive analysis. Predictive analytics like forecast demand, optimize inventory levels, can help to overcome disruptions in the supply chain. Machine learning algorithms can be used to identify patterns and make predictions based on historical data.
5. **Digital Twins:**
   A digital twin represents a physical object or person contextualized in a digital version of its environment. Making digital twins of physical assets, processes, or entire supply chain ecosystems can reduce the costs. Simulate different scenarios using digital twins to assess the impact of decisions before implementation, optimizing resource allocation and efficiency.
6. **Cloud Computing:**
   Cloud computing helps users to use facilities and applications through the internet, without installing and maintaining them in their home or work. Using cloud computing platforms, one can store, process, and analyze vast amounts of data. This can be achieved by real-time collaboration and data sharing across the supply chain.
7. **Artificial Intelligence (AI):**
   AI algorithms are used to predict demand forecasting, route optimization, and predictive maintenance. AI-driven insights can be used to identify areas for improvement in sustainability, circularity, and resource efficiency.
8. **Augmented Reality (AR) and Virtual Reality (VR):**
   AR and VR technologies are used for training, maintenance, and inspection processes in the supply chain. It will help to enhance decision-making by providing immersive experiences for analyzing complex data and visualizing potential scenarios.
9. **Digital Supply Chain Platforms:**
   Digital supply chain platforms can be used to consolidate data and provide a unified view of the entire supply chain. It will enable seamless communication and collaboration among different supply chain partners.

8.8 Case Studies in Circular Supply Chain Decisions

Several case studies demonstrate the successful integration of technology in circular supply chain decisions, showcasing how organizations leverage technological innovations for sustainability, resource efficiency, and circular economy practices. Here are a few examples:

1. **IBM and Plastic Bank:**
   IBM partnered with Plastic Bank to create a blockchain-based solution that encourages the collection of plastic waste in developing countries. Blockchain was used to create a transparent and traceable system, allowing individuals to exchange collected plastic for digital tokens. This initiative supports circularity by incentivizing plastic recycling and preventing ocean pollution.
2. **Provenance and Plastic Packaging Transparency:**
   A UK-based company, collaborated with the Ellen MacArthur Foundation to enhance transparency in plastic packaging supply chains. Provenance uses blockchain to trace the supply chain journey of plastic materials. This transparency helps brands and consumers make informed decisions regarding sustainable and circular packaging choices.
3. **H&M's Circular Supply Chain with AI:**
   H&M, a global fashion retailer, aims to transition towards a circular business model by 2030. H&M is incorporating artificial intelligence (AI) to optimize product design, material selection, and recycling processes. AI helps analyze data on customer preferences, supply chain efficiency, and recycling techniques, contributing to circular fashion practices.
4. **Dell's Closed-Loop Recycling Program:**
   Dell, a technology company, implemented a closed-loop recycling program to recover valuable materials from end-of-life electronics. Dell uses advanced recycling technologies to disassemble and extract materials from returned products. The recovered materials, including plastics and metals, are then used to manufacture new products, contributing to a circular supply chain.
5. **Circular Economy with Cisco's Extended Producer Responsibility:**
   Cisco, a networking and technology solutions provider, emphasizes circular economy principles in its product lifecycle. They use technology to implement Extended Producer Responsibility (EPR) programs. This involves tracking and managing products throughout their life cycle, enabling the company to take responsibility for the recovery and recycling of end-of-life products.
6. **Google's Circular Economy Initiatives:**
   Google has committed to operating its data centers with a circular approach by optimizing resource use and minimizing waste. Google utilizes advanced analytics and machine learning algorithms to optimize data center operations, improve energy efficiency, and reduce environmental impact. The company also invests in renewable energy sources to support circular energy practices.

7. **Nestlé's Blockchain for Transparency:**
   Nestlé, a food and beverage company, aims to enhance transparency and traceability in its supply chain. It employs blockchain technology to trace the supply chain of its products, including coffee and palm oil. This ensures visibility into sourcing practices, supports sustainable production, and enables informed circular supply chain decisions. These case studies illustrate how organizations leverage technologies such as blockchain, AI, IoT, and data analytics to drive circular supply chain decisions. By incorporating these technological solutions, companies can enhance transparency, optimize processes, and contribute to a more sustainable and circular economy.

8.9 Future Directions in Decision Making for Circular Supply Chain

The future of decision-making in circular supply chains will be characterized by the integration of advanced technologies, a consumer-driven shift towards sustainability, and the development of comprehensive circular economy frameworks. Companies that proactively embrace these trends are likely to gain a competitive advantage by fostering resilience, reducing environmental impact, and meeting the evolving expectations of stakeholders.

8.9.1 Emerging Trends and Advancements in Decision-Making Methodologies

As decision-making methodologies continue to evolve, several emerging trends and advancements are shaping the landscape. These trends leverage technology, data analytics, and a more holistic approach to decision-making. Here are some key emerging trends in decision-making methodologies:

1. **Data-Driven Decision-Making:**
   The increasing availability of big data and improved analytics capabilities. Organizations rely more on data-driven insights to inform decisions, optimize processes, and identify opportunities for improvement.
2. **Artificial Intelligence (AI) and Machine Learning (ML):**
   The integration of AI and ML algorithms can do predictive analytics and decision support. AI and ML enables organizations to analyze complex datasets, identify patterns, and make predictions, enhancing decision-making in various domains.

3. **Prescriptive Analytics:**
   Prescriptive analytics suggests actions to optimize outcomes, providing decision-makers with actionable recommendations based on analysis. Beyond descriptive and predictive analytics, is the rise of prescriptive analytics.
4. **Blockchain for Decision Transparency:**
   The blockchain is used to enhance transparency and traceability in decision processes. This helps to improve accountability, reduce fraud, and provide a secure and transparent record of decision-making steps.
5. **Risk Quantification and Management:**
   Improved methodologies are used for quantifying and managing risks associated with decisions. Organizations are better equipped to assess and mitigate risks, leading to more informed and resilient decision-making.
6. **Open-Source Decision-Making:**
   Open-source decision making encourages collaboration, transparency, and contributions from a diverse range of stakeholders in decision-making. These emerging trends reflect a broader shift towards more dynamic, data-driven, and human-centric decision-making approaches. As technology continues to advance, decision-makers have the opportunity to leverage these trends for more effective and resilient decision-making processes.

8.10 Conclusion

Effective decision-making within a circular supply chain is essential for promoting sustainability, reducing waste, and optimizing resource use. This study emphasizes that circular supply chain management (CSCM) hinges on well-structured decision-making models that consider factors such as product design, supplier selection, logistics, and reverse logistics. By incorporating life cycle assessments (LCA), organizations can make data-driven decisions that not only improve operational efficiency but also enhance environmental sustainability. For instance, choosing recyclable materials during product design or opting for sustainable suppliers significantly impacts a company's ability to close the loop in its supply chain. Moreover, the integration of digital technologies such as IoT, blockchain, and AI plays a vital role in enabling real-time monitoring and predictive analytics, which further supports informed decision-making. These technologies help track products across their life cycle, ensuring that resources are reused or recycled efficiently. Additionally, reverse logistics, which involves managing product returns and refurbishments, is critical to creating closed-loop systems that reduce reliance on raw materials and minimize waste. Collaboration across the supply chain—between suppliers, customers, and regulatory bodies—also emerges as a key factor in driving circular practices. Engaging multiple stakeholders helps streamline processes like recycling, product refurbishment, and compliance with environmental regulations. In conclusion, by adopting these decision-making models, companies can not only improve their supply chain's sustainability but also ensure long-term resilience and competitiveness in a rapidly changing global economy. This holistic approach ultimately contributes to the broader goal of transitioning from a linear to a circular economy, where resources are continuously repurposed, fostering a more sustainable future.

References

Bals L, Tate WL, Ellram LM (2022) Circular economy supply chains: from chains to systems. Emerald Publishing Limited, Bingley

Charter M (2018) Designing for the circular economy. Routledge, Abingdon

De Angelis R, Howard M, Miemczyk J (2018) Supply chain management and the circular economy: towards the circular supply chain. Prod Plan Control 29(6):425–437

Dwivedi A, Agrawal D, Jha A, Mathiyazhagan K (2023) Studying the interactions among Industry 5.0 and circular supply chain: towards attaining sustainable development. Comput Ind Eng 176:108927

Farooque M, Zhang A, Thürer M, Qu T, Huisingh D (2019) Circular supply chain management: a definition and structured. J Clean Prod 228:882–900

Khan SAR, Razzaq A, Yu Z, Miller S (2021) Industry 4.0 and circular economy practices: a new era business strategy for environmental sustainability. Bus Strat Environ 30(8):4001–4014

Kilic HS, Kalender ZT, Solmaz B, Iseri D (2023) A two-stage MCDM model for reverse logistics network design of waste batteries in Turkey. Appl Soft Comp:1568–4946

Lehmacher (2017) The global supply chain. Springer

Liu L, Ramakrishna S (2021) An introduction to circular economy. Springer, Singapore

Ncube A, Fiorentino G, Panfilo C, De Falco M, Ulgiati S (2022) Circular economy paths in the olive oil industry: a life cycle assessment look into environmental performance and benefits. Int J Life Cycle Assess 1:1–21

Orji IJ, Ojadi F (2023) The circular supply chain. CRC Press, Boca Raton

Pattnaik S, Nayak MM, Abbate S, Centobelli P (2021) Recent trends in sustainable inventory models: a literature review. Sustainability 13(21):11756

Pruhs A, Kusch A, Woidasky J, Viere T (2023) Design for circularity in manufacturing industries–operationalisation and decision support. Resour Conserv Recycl 202:107376

Ramakrishna Y, Wahab SN (2023) Designing sustainable supply chains to achieve a circular economy. IGI Global

Reslan M, Last N, Mathur N, Morris KC, Ferrero V (2022) Circular economy: a product life cycle perspective on engineering and manufacturing practices. Procedia CIRP 105:851–858

Schröder P (2019) The circular economy and the global south: sustainable lifestyles and green industrial development. Routledge, Abingdon

Settembre-Blundo D, González-Sánchez R, Medina-Salgado S, García-Muiña FE (2021) Flexibility and resilience in corporate decision making: a new sustainability-based risk management system in uncertain times. Glob J Flex Syst Manag 22:107–132

Srivastava B, Sowmya S (2023) Designing sustainable supply chains in india to create a circular economy. IGI Global

Subal L, Braunschweig A, Hellweg S (2023) The relevance of life cycle assessment to decision-making in companies and public authorities. J Clean Prod 435:140520

Taherdoost H, Brard A (2019) Analyzing the process of supplier selection criteria and methods. Procedia Manuf 32:2351–9789

Weetman C (2016) A circular economy handbook for business and supply chains. Kogan Page, London

Chapter 9
Role of Industry 4.0 in the Future Circular Supply Chain Management

Abstract The integration of Industry 4.0 (I4.0) with the circular economy (CE) presents significant opportunities for creating sustainable, resource-efficient supply chains. I4.0, powered by technologies such as the Internet of Things (IoT), big data, artificial intelligence, robotics, and additive manufacturing, enhances operational efficiency, enabling businesses to manage resource use dynamically, reduce waste, and extend product life cycles. The circular economy aims to minimize environmental impact by promoting restorative processes, eliminating waste, and reducing reliance on finite resources. By combining I4.0 with CE, industries can transform traditional linear supply chains into circular systems where materials and resources are continuously reused and regenerated. This paper explores how emerging I4.0 technologies drive the circular economy by enhancing transparency, optimizing production, and enabling predictive maintenance. While the benefits of this convergence are promising, the paper also addresses the challenges of implementation, including financial investment, workforce skills gaps, and technological infrastructure. Ultimately, the integration of I4.0 with CE holds the potential to revolutionize future supply chains, aligning them with sustainability goals.

Keywords Industry 4.0 · Circular economy · Sustainability · Internet of Things (IoT) · Additive manufacturing · Predictive maintenance

9.1 Introduction

The two most significant industrial paradigms that have influenced academia and business in recent years are the circular economy (CE) and industry 4.0 (I4.0). By intention and design, the CE is a restorative or regenerative industrial system. This idea aims to eliminate waste through the superior design of materials, products, systems, and within this, business models. It also replaces the "end-of-life" concept with restoration, shifts to the use of renewable energy, and doesn't use toxic chemicals, which hinder reuse (de Mattos Nascimento et al. 2024). By giving businesses new opportunities to create value, generate revenue, cut costs, be resilient, and

J. Kandasamy et al., *Circular Supply Chain Management*, Circular Economy and Sustainability, https://doi.org/10.1007/978-3-031-81750-2_9

establish legitimacy, the CE enables the decoupling of economic growth from finite resource constraints. Rather, I4.0 is a paradigm that encompasses a broad range of ideas that are impossible to precisely distinguish or classify. The primary driver of I4.0, according to most definitions, is advanced digital technologies. The Internet of Things (IoT), cloud, fog, and edge technologies, blockchain and cyber-security, additive manufacturing, simulation, augmented and virtual reality, Internet of Things (IoT), and autonomous robots and vehicles are the nine technologies that the Boston Consulting Group specifically identified as I4.0's building blocks (Gebhardt et al. 2022). These technologies' integration in an industrial setting can lead to a number of significant increases in competitiveness, including:

1. Technologies of production
2. The state of finances
3. Product life cycle management; market expansion; supply chain management; employee empowerment; and business models.

Considering the prominence these two paradigms have gained over time, a large body of literature addresses CE and I4.0 from a variety of angles. Still, there is a significant gap that exists between theory and reality.

9.2 Technologies for Industry 4.0

Industry 4.0, also known as digital production or smart industry, has emerged over the manufacturing sector. The various kinds of approaches that are available are integrated into a process to address and optimise it. Lean principles have been applied in the past to optimise the system and have consistently advocated for continuous improvement over all organizational models, implying that there is always a window of opportunity for improvement. Within the manufacturing industry, it is widely acknowledged and accepted. The primary goal is to incorporate people into the manufacturing process while maintaining a continuous improvement mindset and raising value-added activities by reducing or eliminating waste. Industry 4.0, also known as I4.0, is the fourth industrial revolution that is sweeping through the manufacturing sectors (Bag et al. 2024). It involves building a smart network that integrates information technology, people, products, machines, and other components with the system's value chain to create smart factories. Thus, both approaches demonstrate ongoing development and improvement while enhancing the system's value chain and achieving operational excellence. This demonstrates the relationship between Industry 4.0 and Lean tools in achieving operational excellence and lowering waste in the system. The manufacturing sector has chosen for many years to apply lean principles and methods to cut down on non-value-added work, simplify operations, and boost productivity. Standardizing procedures, encouraging continuous improvement, and enhancing worker resources on the shop floor are the cornerstones of the lean concepts (Faisal 2023). The complexity of operations has grown dramatically as a result of rising customer and supply demand across the globe, meaning that lean principles by themselves are unable to address operational efficiency. The fourth industrial revolution paved the way for Industry 4.0, which takes a different tack in addressing growing complexity and enhancing output. These methods focus mostly on digital technologies and evaluate each process parameter in order to identify the underlying factors influencing a process's operational excellence. Although the paths and strategies taken to reach the same objective—operational excellence—Lean and Industry 4.0 are not the same (Kumar et al. 2024a). The goal of the lean management approach is to cut waste and non-value-added activities across the system's value chain and process in order to simplify operations and save costs. This method involves the staff in ongoing evaluation in order to increase productivity. Cyber-physical systems, robotics, augmented reality, virtual manufacturing, big data, blockchain, cloud computing, Internet of things, additive manufacturing, simulations, digital twins, optimization algorithms, sensors and actuators, machine learning, AGVs, video and 3D models, human-computer interaction, and horizontal and vertical integration are some of the technologies driving Industry 4.0. Along the entire process or value chain that spans beyond a single organization, these various technologies—sensors, machines, tools, products, and information technology—are integrated. The integrated can communicate with one another, begin analyzing the data collected to identify or anticipate failures, reconfigure the system, and get it ready to adjust to new circumstances. The industries will be able to reach new heights of operational efficiency as a result. For instance, by using the data supplied, maintenance activities can progress from preventive maintenance to predictive maintenance, which involves taking the required actions and actions to reduce the likelihood that a machine will fail, which has an immediate impact on production output (Bag et al. 2024). This enables the teams to have access to the required process data in real time and to be ready for any anomalies. Consequently, this leads to increased system flexibility and overall efficiency gains. Industries have been trying to figure out how to combine Industry 4.0 and traditional lean concepts in order to maximize efficiency in recent years. Numerous research projects are underway regarding the integration of both approaches, either in parallel or consecutive order. This paper aims to identify various types of lean tools, provide appropriate definitions, and discuss how industry 4.0 technologies can improve lean tools. The industries can integrate the application of lean management and I4.0 to achieve the next level of operational excellence thanks to the relationship that has been established between the lean tools and I4.0 techniques (Junaid et al. 2024). The goal is to identify lean and green concepts and their unique strengths for process optimization within a system. A variety of Industry 4.0 technologies are also examined, along with their potential to support the system's seamless technological transformation. Although the current lean and green tools are highly valued, the main goal of this study is to find a common best suit relationship between the lean-green concepts and the state-of-the-art industry 4.0 technologies, so that the industries can effectively and efficiently adapt to this dire situation.

9.3 The Meaning and Development of Industry 4.0

The Industrial Revolution began with mechanization, then moved on to steam power, weaving, mass manufacturing, assembly line development, and the use of electrical energy. Subsequently, the procedure is made more precise and accurate by automating it and storing the data with the use of electronics and computers (Kayikci et al. 2024). These days, technology advancements are connected via networks to optimise the entire organization, not just the process. It applies the ideas of a cyber-physical system using the internet and cutting-edge technology to enhance human-machine interaction. This results in the industry being mass customized and all entities being integrated into the system's value stream. The intelligent factory is made possible by the Internet of Things, which facilitates system integration across a network. End-to-end information and communication can be easily flown across the whole system thanks to the automation of warehouses and manufacturing facilities and digitally produced devices that can communicate with other machines. This outlines how to connect the supply chain, warehousing, marketing, logistics, and consumer demand with the production system to improve organizational visibility.

Industry 4.0 has a large influence on process execution and has a stronger impact than the manufacturing industry. Because of the real-time data transfer and information flow from end to end of production planning for vigorous self-optimization, the system would provide a better picture of forecasting than the traditional method (Karmaker et al. 2023). The development in automation on the factory floor is a result of the advent of information and communication systems via industrial networks. The creation of intelligent or smart manufacturing systems through the integration of various technologies allows us to simulate the entire value chain, including inventory, logistics, production transport, warehousing, and product usage, all of which have a positive effect on the production floor. This process starts with the ordering of materials from suppliers and ends with customer delivery with various industry 4.0 tools as shown in Fig. 9.1 and they are incorporated at different levels of the organization.

The term "concept" is used to refer to a broad range of emerging technologies, including robots, blockchain, 3D printing, virtual manufacturing, augmented reality, Internet of things, and cyber-physical systems (Kiran and Sahin 2024). The development of an intelligent or smart system, which consists of smart equipment, goods, and gadgets with the ability to communicate and exchange information to initiate the required actions and exert control over one another, depends heavily on the acceptance of these technologies. A complicated technology system called Industry 4.0 is being studied by a number of academics and researchers. With a paradigm change in the way work culture is organised, corporate strategies, and manufacturing technologies are used, it has enormous potential for the industrial and production sectors, offering several opportunities in the social and economic spheres (Kumar et al. 2024a).

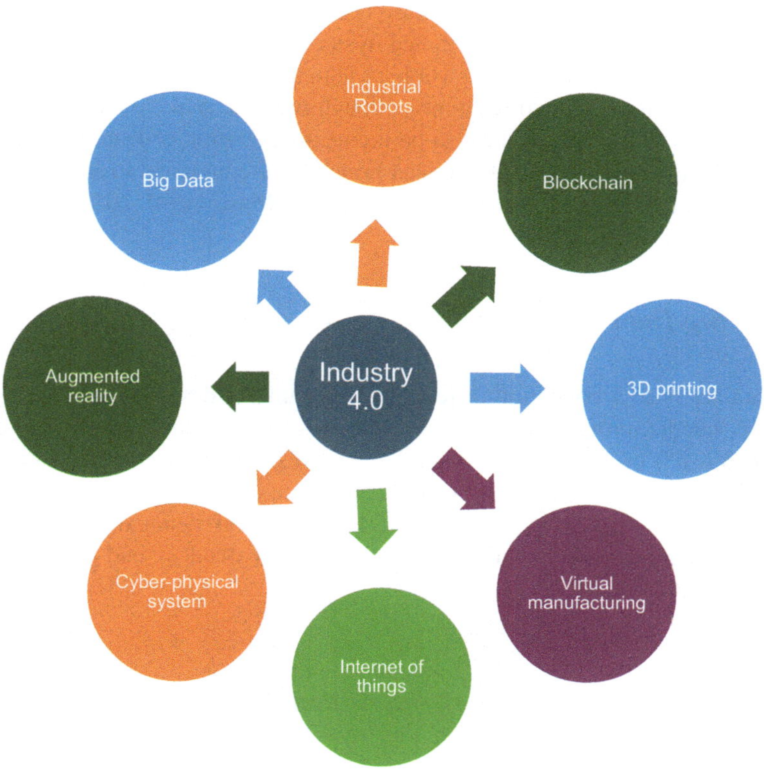

Fig. 9.1 Industry 4.0 technologies

9.4 Investigating How Emerging Technologies Are Influencing the Future

The fourth industrial revolution in the manufacturing sector is being driven by technological advancements such as cyber-physical systems, the internet of things, autonomous robots, cloud computing, big data, analytics, 3D printing, augmented reality, cyber security, and others. At the moment, information is shared at the device level both by absorbing data transmitted by other devices and by detecting and transmitting the surroundings (Kumar et al. 2024b). Additionally, because the sensors are lightweight and compact, they are deployed more often. A new decision model is developed based on the new parameter inputs after the data analysis with the modified parameter from the new scenario is taken into account. Applying industry 4.0, which is the visual component that sits atop the architecture and exports the whole virtual representation, data, and intricate analysis to the end user, is crucial. This part is crucial since it simplifies the top layer application and connects with all of the functionality and data structures. Since this is the visualization medium between the user and the environment, it must be user-friendly as the user may not be a technical person. This phase is expressly implemented as it integrates between the application and distributed system. This enables information to be shared throughout the system as a whole or among specialized systems, provides a uniform platform, and identifies trends through big data analysis. Because of this, industry 4.0 is primarily intended to be the next generation of lean concepts. The following outlines the areas that the lean principles are intended to address as well as how industry 4.0 may facilitate these categories (Mastos et al. 2021).

9.4.1 Big Data

Big data refers to a collection of technical instruments designed to gather, store, organise, handle, process, and evaluate vast amounts of data. Big data was developed as a result of the inability of traditional tools to complete these tasks within reasonable time frames. Big data facilitates the use of IoT in the following areas: communications, radio frequency (RFID), supply and demand, manufacturing logistics, radio frequency identification (sensors and barcodes), and product identification (Lu et al. 2024).

9.4.2 Robotics-Based Industrial Automation

Software that incorporates artificial intelligence (AI) is referred to as industrial automation. It can systematically do repetitive operations that were previously completed by people on a big scale thanks to its machine learning capabilities (Lu et al. 2024).

9.4.3 Models and Simulations

Systems taken out of the actual world are modeled using simulations in order to evaluate their behaviour under various circumstances and comprehend how they react to various direct and indirect agents that might affect the system's outcomes. Simulations are an essential component of Industry 4.0 because of their versatility, which allows them to be used in a wide range of industries including manufacturing, services, design, and healthcare (Liu et al. 2023). Software-based process simulation is deemed advantageous by I4.0 due to its ability to implement analytical procedures.

9.4.4 *Systems of Integration*

In order to offer the interfaces needed to integrate virtual and physical system objects, this idea refers to connecting system components (horizontal integration) or two or more systems (vertical integration). Another name for it is end-to-end integration (Liu et al. 2023).

9.4.5 *Internet of Things (IoT)*

The Internet of Things (IoT) was first conceptualized by Kevin Ashton in 1999 as a system that sends data in real time about products, services, processes, activities, or tasks in order to allow dynamic information management. Conversely, the Internet of Things (IoT) is characterized as the network of electronic devices that communicate and gather data in real-time, adding value to the finished product (Lu et al. 2024).

9.4.6 *Protection Against Cyber Attacks*

In order to increase the efficiency of industrial systems, Cybersecurity assists enterprises in developing safe business controls as well as the procedures used to evaluate such controls. Cybersecurity refers to the use of cyber-physically interconnected intelligent systems to automate industrial processes via secure networks and the provision of security standards that provide defence against cyber attacks (Kumar et al. 2024a).

9.4.7 *Utilizing the Cloud for Processing*

A digital infrastructure known as cloud computing makes it easier to access computers and processing engines that are housed on distant servers over the internet (Kumar et al. 2024b). The implementation of electronic commerce services, which facilitate vendor-customer engagement through resource virtualization and the automation of business processes and logistics, is now being pushed through the usage of cloud computing.

9.4.8 Additive Manufacturing

Additive manufacturing differs from traditional manufacturing techniques in that it generates product parts while minimizing waste and eliminating the need to purchase and utilize specialized equipment. Without the need for prefabricated mould, production is based on three-dimensional (3D) digital designs and traditional raw materials. The primary tools used in additive manufacturing are 3D printers (Nishal et al. 2023).

9.4.9 Virtual and Augmented Reality (AR, VR)

I4.0 uses augmented reality (AR), a technology that allows users to access virtual places superimposed on real-world physical spaces using digital tools and virtual information data. While augmented reality (AR) does not replace the actual world, it does provide a means of gathering data from simulations to enhance comprehension of real-world systems. The ability of Industry 4.0 Technologies (I40Ts) to significantly transform industry and the need for this transformation to be monitored from a sustainability science perspective in order to provide solutions to environmental problems in industry are highlighted, along with the transformative potential of these technologies in the context of circular supply chains (Narkhede et al. 2024). This viewpoint is vital to consider since it enables businesses to integrate environmental solutions with overall competitiveness and quality. Nevertheless, it was shown in one study on I40Ts that just 18% considered a sustainability-related viewpoint. Furthermore, there are fewer research on I40Ts that include CE. Furthermore, the majority of research on the impact of I40Ts on CE has addressed the topic in a broad sense; relatively few have contrasted the effects of the various technologies. Business value is being driven by the observable benefits brought about by the deployment of I40Ts. Numerous enhancements are contributing to the decrease in resource use. Businesses may increase labour, capital, material, energy, time, and resource consumption efficiency with the use of I40Ts. Several writers have noted that additive manufacturing (AM) technology has the potential to lower raw material and energy usage when analysing the impact of particular technologies.

9.5 The Concept for a Future of Circular Supply Chains (Nishal et al. 2023)

An industrial system that is restorative or regenerative by design is known as a circular economy. It attempts to eradicate waste and replace the idea of end-of-life with restoration. It also moves away from the use of hazardous chemicals, which hinder reuse and return to the biosphere, and towards the use of renewable energy.

Beyond the take-make-dispose extractive industrial paradigm of today, a circular economy seeks to reinvent growth by emphasising beneficial effects on society as a whole. It means engineering waste out of the system and progressively severing the link between economic activity and the use of limited resources. The circular model, which is based on the switch to renewable energy sources, increases social, natural, and economic capital. It is predicated on three ideas: eliminate waste and pollution, maintain the useful life of goods and resources, and restore natural systems.

9.6 Establishing the Long-Term Objectives of Circular Supply Networks

The fundamental goal of a circular economy model is to eliminate waste. Actually, the foundation of a circular economy is the notion that waste doesn't exist (Ozkan-Ozen et al. 2020). To do this, items are made to last (high-quality materials are utilised) and are optimized for a cycle of disassembly and reuse, which makes handling, transforming, and renewing them easier. Ultimately, the close product cycles set the circular economy model distinct from recycling and disposal, which involve significant labour and embedded energy losses. The ultimate objective is to balance the flows of renewable resources and manage finite stocks in order to protect and improve natural capital. Technical and biological cycles are distinguished in the circular economy paradigm. Only in biological cycles, when biologically-based resources (such food, cloth, or cork) are intended to be recycled back into the system through anaerobic digestion and composting, can consumption take place. Future renewable resource supplies come from these cycles, which rejuvenate living systems like the seas and oil. Optimising resource yields is ultimately one of the goals of the circular economy, which aims to do this by reusing resources, products, and components as long as possible in biological and technical cycles (Shakur et al. 2024). The last tenet of a circular economy is that, in order to reduce reliance on resources and boost system resilience, the energy needed to power this cycle should naturally be renewable. In this way, identifying and eliminating negative externalities helps to increase the efficacy of the systems.

9.7 Circular Approaches Have Long-Term Advantages for the Environment and Businesses (Gebhardt et al. 2022; Liu et al. 2023; Ülkü et al. 2024)

9.7.1 Assist in Cutting Greenhouse Gases

The circular economy aims to create a sustainable model that minimizes waste and environmental impact while maximizing resource efficiency. By shifting from a linear "take-make-dispose" approach to one that emphasizes reuse, recycling, and regeneration, it seeks to prevent overuse of natural resources, reduce greenhouse gas emissions, and protect ecosystems. Key strategies include reducing reliance on fossil fuels through renewable energy, promoting dematerialization and reuse to minimize new material needs, and maximizing the use of leftovers and non-toxic materials. Ultimately, the circular economy offers a pathway to a more resilient and sustainable future by re-imagining how we produce, consume, and manage resources. A shift to a circular economy may lessen demand on virgin materials by 28% and cut greenhouse gas emissions by 39%, according to the most recent Circularity Gap Report. Because less material needs to be mined, processed, and transported throughout product manufacture, the circular economy makes it possible for product carbon footprints to be reduced.

9.7.2 Raise the Rates of Recycling and Reuse

Reusing materials and goods is encouraged by the circular economy, which makes it simpler for businesses to fund recycling initiatives. In turn it allows the valuable usable material to recycled, instead of reaching the landfill which increases the recycling rate. According to a research by the Ellen MacArthur Foundation, there would be a 14% gain in resource productivity, a 12% reduction in energy consumption, and an 11% decrease in carbon emissions for every 1% increase in recycling and reuse. Natural ecosystems like forests, which are home to a rich diversity of living forms with intricate interactions between them and where no portion goes to waste, are examples of this.

9.7.3 Diminish the Harmful Effects of Manufacturing on the Ecosystem

The global challenge of waste management persists as nations generate more garbage than they recycle, leading to the export of undesirable or useless goods to neighboring regions or even within the same nation. Manufacturing, responsible for nearly 70% of global energy consumption, stands as a significant contributor to pollution and climate change. Its impact is evident in the declining air quality experienced in many regions worldwide. Addressing the environmental consequences of manufacturing processes and waste management practices is crucial in mitigating pollution and combating climate change on a global scale. Fortunately, manufacturers may embrace more sustainable practices because the circular economy concept revolves upon the notion of reuse.

9.7.4 Lower Energy Expenses

With the circular economy emphasizing "smart" consumption and increased energy efficiency, energy prices should decrease as green economy technology progresses. Because there is a significant financial incentive for consumers to purchase energy-efficient items, the circular economy promotes this behaviour. For instance, an owner's capacity to repair a washing machine is enhanced if it is made to be simply disassembled.

9.7.5 Reduce Degradation of Land

In order to counteract these losses, the circular economy seeks to boost land productivity, decrease food waste from farm to table, and enhance nutrient return to the soil. In addition to causing biodiversity loss, land degradation drains $40 billion from the US economy annually. Approximately 25% of the world's geographical area has been degraded. Thankfully, producers are beginning to take into account how their products may affect the environment along the whole production chain. The circular economy encourages more sustainable practices by making it simpler for businesses to track their effects on land utilization. Waste is not thrown away but rather treated as a resource in a circular paradigm. Thus, this procedure can support the preservation of land use. Furthermore, it aims to improve the quality of the soil by means of biological processes such as anaerobic digestion and composting, hence minimizing the requirement for extra nutrients and fertilizers.

9.7.6 Encourage Natural Resource Conservation

The circular economy has the ability to save up to 70% of materials, which may add up to a trillion dollars in savings annually, in contrast to the linear economy, which is based on continuous material extraction. Because of this, a circular economic system is effective in meeting the population's growing material demands by prolonging material cycles, increasing their lifespan, minimizing landfills, and using energy-based recycling techniques. Additionally, it reduces the pollution that comes from the mining of materials.

9.7.7 Create a New Company Out of Rubbish

As nations experience economic growth and rising living standards, the accumulation of waste from both consumers and industries becomes a significant challenge. In many cities, solid waste management consumes a sizable portion of the budget. Nevertheless, forward-thinking businesses are leveraging circular economy concepts to convert waste into profitable ventures. By aggregating sufficient volumes, these companies can invest in the infrastructure necessary to efficiently manage and organize trash supply chains, thus not only addressing waste management issues but also creating new sources of revenue and promoting sustainable practices. Three instances from McKinsey's study illustrate this: although polyethylene terephthalate bottles with mixed waste may be burned, not much energy is generated. However, a far higher return is produced when the material value of polyethylene terephthalate bottles is extracted from mixed recyclables or bottle-to-bottle recycling. One of the most common materials to recycle is tyres. Conversely, metals are typically extracted from car tyres during backyard fires that are open, which has detrimental effects on the environment and human health. Tyres may be broken down into aggregate that can be used as road paving material, increasing its value by nearly 10 times when combined for use as industrial fuel. Yields are increased by using liquid-chemical extraction methods or the finest smelting techniques for electronic trash instead of small-scale recycling. Remember that, pound for pound, electronic scrap contains more gold than ore.

9.7.8 Create New Prospects

It is crucial to recognize that resource consumption and economic growth are two completely distinct things. When the sole means of achieving success is the use of resources, the economy cannot expand. But it does so effectively by opening up fresh possibilities for material recycling. According to a significant research conducted with the Ellen MacArthur Foundation, this strategy might increase Europe's resource productivity by 3% by 2030, saving €600 billion in annual costs and yielding additional economic benefits of €1.8 trillion.

9.7.9 Encourage Entrepreneurship and Innovation

In the realm of circular economics, novel perspectives prompt innovative solutions to economic challenges. Emerging technologies offer businesses fresh avenues to extract value from resources. Consequently, developing nations are investing in research and development (R&D) to explore these opportunities further. One notable trend is the adoption of biotechnologies, which harness biological processes to create novel materials and products. For instance, agricultural waste can be transformed into biodegradable packaging materials through bioengineering techniques. This example showcases how the circular economy fosters creativity and drives technological advancements that benefit both the economy and the environment. In order to provide environmentally friendly packaging choices, Casper Material is creating mushrooms to replace plastic. The sharing economy and assurances of origin are two examples of the new business models that will result from all these opportunities. This calls for optimizing the system as a whole and taking into account circular value chains as opposed to linear ones. This results in fresh perspectives, interdisciplinary cooperation between manufacturers, recyclers, and designers, as well as environmentally friendly inventions.

9.7.10 Provide Additional Job Opportunities

Job generation is another major benefit for the client. Many new sectors will be required in order to achieve a circular economy. This suggests that a large number of new job possibilities will become accessible. However, a significant global concern is that many environmental strategies may remove some employment opportunities, such coal mining or other jobs involving non-renewable resources. Employment growth of up to 9% is expected in the recycling, remanufacturing, and repair industries, which will help facilitate the shift to a circular economy. Labor-intensive recycling and high-quality repairs, as well as jobs in the logistics industry from local product take-back and new company ventures through innovation and the service economy and new business models, will all contribute to the expansion of these occupations.

9.7.11 Increase the Resilience of Businesses

A circular economy model is immune to the fluctuations in the price of raw materials since it is predicated on the reuse and recycling of commodities. By helping companies maintain their supply networks, it enables them to continue operating. It also equips businesses to handle unanticipated circumstances. A business can carry on even in the event that adverse weather prevents them from obtaining the necessary supplies or from producing their products. They can turn to alternative material sources and substitute recycled materials for regular ones used in manufacturing.

9.7.12 Offer Advantages to the Customer

Customers of today like to support businesses that share their values, and one of their top priorities is supporting green efforts. Using an environmentally friendly company approach might help you increase the number of devoted clients you have. There are several advantages for consumers in a circular economy in addition to environmental ones. Reusing materials, for instance, prevents planned obsolescence, extending the lifespan of your goods. Since it promotes frugal behavior like leasing or renting rather than buying, buying second-hand goods, and other such actions, it also promises a rise in disposable income.

9.7.13 A Rise in Service Demand

The need for these services is growing as more businesses embrace the circular economy. A policy like this fosters sustainable development, which is good for the economy as a whole. According to a report by the Ellen McArthur Foundation, a circular economy can generate demand for new services and job opportunities. Reverse logistics companies work to reintroduce products into the system. Selling and marketing platforms strive to provide long-lasting products. Companies that specialize in product refurbishing and remanufacturing parts and components. Businesses that lease or rent things rather than selling them are supported by the circular economy. Businesses that use such a system frequently communicate with their consumers, which gives them valuable insights about the consumption habits and preferences of their clientele. In the end, this can increase user loyalty and satisfaction. Additionally, the company may assist it in creating goods and services that better fit the needs of its clients and adhere to the circular economy's philosophy of producing long-lasting goods in response to market demands.

9.7.14 Encourage Cooperation

A number of sectors, including the government, legislators, manufacturers and consumers, investors, innovators, and financial institutions, must get involved in the circular economy. It's a big job that needs these important people working together to succeed. Contributions from the north to the south at all societal and industrial levels are also included in this. It is a potent model that makes it possible for everyone to help it succeed. In this system, everyone has a part to play in ensuring that resources are used regularly and nothing is wasted.

9.8 The Significance of Sustainability as a Catalyst for Upcoming Supply Chain Tactics

Because of the existing linear economy, our natural resources are being severely exploited to meet the needs of a growing population. As the circular economy develops, we will have the ability to create models that protect the environment and natural resources while meeting the demands of an expanding population (Ülkü et al. 2024). The difficulty we have today is mainstreaming and scaling up the circular economic model, even though it has begun to find its way into the linear economic system. But given the present upheaval, it's imperative that we embrace the sustainable economic model of the circular economy. Although it takes time, in the long run the positive externalities that it ensures are significantly more significant. Begin by determining how to live up to the core values by posing queries such as,

1. How can we create designs with waste?
2. How could materials be reintroduced into the system?
3. How can natural resources be replenished?

A circular economy will need to be established globally, requiring the cooperation of enterprises, governments, and citizens. A wide range of business sectors are seeing the impact of the circular economy on the development of more sustainable enterprises. Sustainable business research is a popular topic. While this influence is more obvious in certain places and is perceived as a fundamental problem, the true effects of circular economy strategies on enterprises are not always clear. However, even if many have been in use for a long time, many activities or behaviours that promote circularity—or more circularity—might not have been recognised as such and have instead been given different titles. Circular economy can push businesses and organizations towards more sustainable practices in a number of areas, including research and development, logistics and reverse logistics, quality control, environmental management, cost management, service management, and strategic planning. These activities can include direct, unidirectional impact from one over another as well as reciprocal, multidirectional changes within the sectors and others.

9.9 Industry 4.0 and Circular Supply Chain Intersection (Veile et al. 2024)

"Are contenders to be two sides of the same coin" are I4.0 and CE. The establishment of a new technological evolution and the development of a circular paradigm are related in that, on the one hand, CE is meant to be self-regenerating; on the other hand, the biological cycle needs to return to the biosphere, and the technical cycle needs to be managed so that it can renter the environment with the least amount of damage. On the other hand, I4.0 will promote the introduction of new Business models (BMs) that are also focused on environmental and social aspects, aiming to improve the well-being of people and the environment, as a result of the increased interconnection of businesses and resources. Therefore, the adoption of a circular paradigm is closely tied to recent technological advancements. Digitization can facilitate the shift towards BM that is more sustainable by helping to close material loops by supplying data on the availability, location, and state of goods. Additionally, it can assist in minimizing transaction costs and waste. Producers can monitor and optimise the performance of their products with smart and connected products. Predictive maintenance can also be implemented, which can increase product reliability and potentially extend its lifetime. Service tasks and travel routes are optimized, the supply chain (SC) will be more synchronized, and the energy consumption will be optimized with consideration for the environment, all thanks to data and algorithms. In the context of Circular economy (CE), IoT can link stakeholders across the SCs by using data from sensors; these connections could serve as dynamic feedback control loops. The ability to extend product lifecycle management beyond the boundaries of producers is made possible by interconnected SCs. However, big data (BD) analytics is thought to be a useful method for providing context for the vast amount of data captured by IoT, facilitating improved decision-making. Out of all the I4.0 technologies currently in use, cloud manufacturing (CM) and additive manufacturing (AM) show great promise for a practical application of circular economy principles in the manufacturing setting.

Cloud manufacturing and cloud computing are viewed by businesses as efficient ways to complete tasks quickly and cheaply. In addition to cutting expenses, cloud computing has a positive environmental externality because resources are pooled and effectively utilised by multiple businesses. Businesses that shift to cloud computing will be able to drastically cut down on their energy use and carbon emissions; virtualized servers and data centres should be used in place of physical ones, and cloud infrastructures can be hired for processing power. The reduction of the environmental footprint of cloud computing is primarily driven by four factors: data centre efficiency, server utilization, multi-tenancy, and dynamic provision. Actually, because of cloud dynamic allocation, capacity is actually managed more carefully, which lowers inefficiencies. Additionally, the cloud providers minimize their environmental impact by serving multiple businesses concurrently on the same server infrastructure, sharing instances amongst them. Higher performance and utilization would result from a higher rate of server infrastructure utilization, much like the multi-tenancy feature. Lastly, data centers have equipment and infrastructure that is designed to be as innovative and environmentally friendly as possible, and they are optimized for power efficiency. However, because of its great efficacy and efficiency, additive manufacturing, or AM, must be acknowledged as a major factor in the shift to a circular economy. The custom single part production principle that AM technologies generally adhere to circumvents traditional design guidelines. In contrast to the idea of economy-of-scale, this technology makes the concept of economy-of-one feasible. This technology reduces emissions, wastes resources, and produces complex parts more effectively. In addition to minimizing waste, the precise addition of material results in a smaller environmental impact.

9.10 Future-Focused Goals That Industry 4.0 and Circular Supply Chains Have in Common

This new technology is not limited to the manufacturing sector; it can be used in any industry that produces a wide variety of products, including the food, beverage, and automotive sectors. Additionally benefited are industries like pharmaceuticals and semi-conductors where high quality and precision are required. Companies must choose the right infrastructure in order to benefit from the latest technological developments in industry 4.0 (Faisal 2023). The industrial revolution's technological advancements significantly boost productivity in response to client demands. It establishes the framework for the introduction of fresh, creative business models that satisfy a range of human needs. In order to experience rapid development and revenue, organizations must introduce new, innovative products with digital capabilities. Smart services based on business analytics and available on the internet will boost industry growth. Industry 4.0's power lies in its ability to exchange data with new technologies in a seamless manner, adding value throughout the ecosystem. Important developments in Industry 4.0 indicate a discernible shift in the market's dynamics from digital high-tech to industrial equipment.

The increased demand for industrial systems in a growing economy presents a challenge, as companies must prioritise the sustainability of their products. The emergence of Industry 4.0 presents immense potential for achieving sustainable manufacturing. Organisations can increase the lifespan of their equipment by researching the environmentally beneficial aspects of sustainable product manufacturing and utilising CPS and retrofitting techniques. According to Industry 4.0's fundamentals of sustainability, businesses should implement closed-loop product life cycles (Karmaker et al. 2023). In order to achieve higher levels of throughput, organizations should take advantage of the opportunities to enable methods for reusing and remanufacturing the products. The goal of zero defects in wheel machining automation is attained through the use of Automatic Virtual Metrology in conjunction with a new platform known as the Advanced Manufacturing Cloud of Things. This opens the door to the next stage of Industry 4.0. Industry 4.0 will only come to pass when businesses are prepared to move towards digital technologies. Conventional supply chains have a great deal of potential to become highly efficient digital supply chains with the help of Industry 4.0 by cleverly connecting suppliers, customers, manufacturing, logistics, product development, and procurement (Kiran and Sahin 2024). If industry 4.0 can advance in allowing digital supply chain sustainability, the ecosystem as a whole will gain. In an entire digital ecosystem, smart products and services can be provided by disruptive business models to cater to customers. The hub that disseminates information throughout the supply chain and is primarily identified by acting decentralised from a distance using a tightly integrated platform is the digital cyber network.

9.11 Examples of the Proactive Incorporation of these Ideas (Junaid et al. 2024; Karmaker et al. 2023; Kiran and Sahin 2024)

9.11.1 Waste-Related Internet (IoW)

Closing the loop after a product's use phase is over is one of the circular economy's challenges. Opportunities exist to improve the post-use collection process' effectiveness, efficiency, and sustainability with I4.0 technologies. AllWaste, an Internet of Things (IoT) waste collection and street cleaning solution in Santander, Spain, serves as an example of this. The system monitors and notifies field operators and supervisors of container fill levels through the use of telemetry equipment and ultrasonic sensors. As a result, collection frequency and routes can be planned more effectively using real-time, geolocated waste generation data. Containers can be uniquely tagged, and collectors can verify activities with supervisors by using handheld RFID (radio-frequency identification) readers to identify containers. The efficient routes designed by All Waste cut down on fuel use and greenhouse gas emissions, enhance inventory management, and lessen olfactory, visual, and noise pollution. In order to maximise event resolution and minimise response times, the system also features a mobile application that enables residents to report incidents to the street cleaning and collection service.

9.11.2 Blockchain for Food Industry Circularity

The traceability of products over their entire life cycle is another essential component of the circular economy. In order to meet the demands of an increasing number of stakeholders for transparency, to comply with laws and standards, and to meet the operational needs of supply chain actors, information must be gathered at each stage of the cycle. Comparing blockchain technology to traditional technology reveals several advantages in terms of automation, information encryption, security, and communication between chain participants who utilise various computer systems. Based on blockchain technology, the IoTrace tool tracks component and shipment locations across the supply chain ecosystem and incorporates extra data about the licences and certifications of different chain actors. It also makes it easier for a company's partners to exchange logistics data. This turns sustainability and circularity into a competitive advantage by showcasing and strengthening them. IoTrace is used by a global fishing company to integrate third-party records and monitor fishing techniques, landing procedures, and other value chain activities. Following that, customers can access all of this data by clicking on links contained in the lot numbers on product packaging.

9.11.3 IoT and Blockchain to Circularise the Textile Sector

IoTrace is also being used to assist the textile industry in closing the materials loop. Certain traceability requirements about the materials obtained, the transformations performed, and the finished product sent to customers are implied by the use of recycled textiles. All product traceability data in the intermediate phases of the textile reuse process is digitalized by the IoTrace application. Complete transparency is made possible by the fact that this data is visible to all parties involved and is irreversibly linked in the blockchain network. Users may choose to add traceable data to the blockchain network directly, through third-party systems, or through Internet of Things (IoT) devices such as sensors.

9.12 Progress in IoT and Real-Time Monitoring

Supply chain IoT provides never-before-seen real-time visibility throughout the whole chain. Incorporating sensors, devices, and data analytics allows companies to track goods, materials, and shipments at all stages of the supply chain journey. Increased visibility dramatically reduces uncertainty and increases transparency, two factors that are frequently problematic for traditional supply chain operations.

9.12.1 Monitoring Goods and Shipments

IoT-capable devices are intelligently placed on products and shipments. These sensors collect data on humidity, temperature, location, and other relevant characteristics. A central platform receives real-time data, allowing beneficiaries to monitor the status and state of commodities as they travel via the supply chain network.

9.12.2 Enhanced Transparency

Supply chain managers can instantly learn about the status of products thanks to the Internet of Things. They are able to locate shipments precisely, anticipate possible delays, and take preventative action to deal with any problems. Since accurate information is easily accessible, transparency fosters trust with partners and customers.

9.12.3 Decreased Uncertainty

Because of limited visibility, traditional supply chain management frequently entails some degree of uncertainty. This is lessened by IoT, which offers constant updates on the location and state of goods. This lowers the possibility of expensive mistakes and delays by enabling decision makers to make well-informed decisions based on real-time data.

9.12.4 Inventory Management

IoT in the supply chain transforms the monitoring and management of inventory levels, which has a big impact on inventory control. RFID (Radio Frequency Identification) tags and connected sensors are essential for inventory optimisation.

9.12.5 Constant Monitoring

IoT-capable sensors are integrated into inventory items and storage spaces. These sensors keep an eye on expiration dates, temperature, and stock levels all the time.

9.12.6 Automated Alerts

IoT sensors set off automated alerts when inventory levels cross certain thresholds. Supply chain managers receive these alerts and are prompted to take appropriate action, such as restocking orders or modifying production schedules.

9.12.7 Decreased Excess Inventory

When products go out of style, excess inventory can waste money, tie up capital, and raise storage expenses. The predictive analytics and precise monitoring provided by IoT allow businesses to keep their inventory levels at ideal levels. Businesses that order goods based on real use patterns might decrease excess inventory and associated expenses. Preventing stockouts: Running out of stock can lead to missed sales opportunities and disgruntled customers. Real-time IoT monitoring ensures that inventory levels and demand are balanced. Businesses can prevent stockouts, quickly complete customer orders, and uphold a positive brand image with the help of timely alerts and insights.

9.12.8 Analytics for Predictive

In supply chain management, IoT-generated data makes predictive analytics possible. Businesses may uncover trends, obtain important insights into demand patterns, and allocate resources more wisely by utilising both historical and current data.

9.12.9 Historical and Real-Time Data

Internet of Things (IoT) sensors are always gathering data on different parts of the supply chain, such as orders from customers, production rates, transit duration's, and stock levels.

9.12.10 Forecasting Demand

Predictive analytics models are able to predict demand more precisely because they have access to a multitude of data. By examining past sales data and taking current market trends into account, businesses can predict future demand trends.

9.12.11 Trend Identification

Supply chain managers can spot new patterns and trends by using data generated by the Internet of Things. This could involve modifications to the dynamics of the market, seasonal variations, or shifts in consumer preferences. Early detection of these trends by predictive analytics algorithms enables businesses to modify their plans and deploy resources appropriately.

9.12.12 *Making Informed Decisions*

Throughout the supply chain, decision-makers can make well-informed decisions when equipped with predictive insights. This could entail fine-tuning distribution routes, optimising inventory levels, or varying production volumes. There is less reliance on conjecture because these choices are grounded in data-driven projections.

9.13 Potential Applications of IoT for Real-Time Supply Chain Visibility in the Future (Gebhardt et al. 2022; Junaid et al. 2024; Kayikci et al. 2024)

9.13.1 *Visibility of the Supply Chain*

IoT technology revolutionizes supply chain management by enhancing end-to-end visibility. With real-time tracking and monitoring capabilities, companies can precisely estimate shipment arrival times and optimize routes, leading to more efficient logistics operations.

9.13.2 *Real-Time Tracking*

IoT-enabled sensors offer ongoing updates on shipment location and condition, facilitating real-time tracking throughout the supply chain journey, from production to delivery. This capability empowers supply chain stakeholders to monitor goods' progress at every stage, enabling proactive management and ensuring timely delivery.

9.13.3 *Accurate Arrival Times*

By harnessing IoT-generated data, companies can compute estimated arrival times with remarkable precision. This data is indispensable for orchestrating downstream processes like inventory management and order fulfillment. Accurate arrival time predictions mitigate uncertainty, enhancing operational planning and ensuring smoother workflows.

9.13.4 Efficient Route Optimization

Through IoT-enabled devices, valuable data on traffic conditions, weather patterns, and road closures is collected in real-time. Supply chain managers leverage this data to optimize routes for delivery trucks, ensuring punctual deliveries while minimizing fuel consumption and transportation expenses. This proactive approach not only enhances operational efficiency but also contributes to sustainability efforts by reducing carbon emissions associated with transportation.

9.13.5 Enhanced Customer Experience

Enhanced visibility afforded by IoT technology enables companies to furnish customers with precise updates on their order statuses. Customers can track their shipments in real-time, fostering greater satisfaction and trust in the company's service. This transparency cultivates positive customer experiences, solidifying loyalty and bolstering the company's reputation.

9.14 Predictions on How IoT Will Contribute to Enhanced Circular Practices (Karmaker et al. 2023; Kumar et al. 2024a)

The convergence of AI and IoT technologies within the circular economy realm presents unprecedented opportunities for value creation. This synergy fosters the emergence of a "factory of the future" model characterized by reduced costs, heightened efficiency, and the feasibility of high-quality, customized production runs. Industries such as construction, food, manufacturing, and recreation are poised for transformation through the integration of emerging technologies like 3D printing, advanced robotics, wearables, and augmented reality into supply chain, logistics, and asset management processes. The embedding of low-cost sensors in reusable products/devices enables efficient asset management within the circular economy. While there may be initial costs associated with IoT implementation, it facilitates processes that minimize waste and promote material recycling. The integration of IoT throughout the value chain enhances efficiency by enabling real-time monitoring and analysis, optimizing resource utilization from production to end-of-life management.

IoT-driven asset retrieval during the product life cycle facilitates effective reverse logistics, thereby facilitating product recycling and extending the lifespan of goods. Furthermore, IoT-enabled leasing models align with circular economy principles, as manufacturers transition from selling to leasing assets. By monitoring asset condition and dynamically repairing them as needed, manufacturers are incentivized to produce durable products, reducing disposability and resource wastage. Digital transformation and big data analytics provide substantial support for circular economy business models. Product-as-a-service models and the sharing economy contribute to dematerialization, significantly impacting sustainability across various industries. The product life cycle is closely monitored to assess reusability potential, with predictive analysis guiding optimization efforts. Design for circular economy strategies are employed to minimize waste and maximize benefits based on the assessed reusability potential. Overall, the integration of IoT, blockchain, AI, and big data enables efficient monitoring and control within the circular economy ecosystem, driving sustainability and innovation forward.

9.15 Anticipated Advancements and Case Studies Demonstrating Future Scenarios (Faisal 2023; Gebhardt et al. 2022)

Data centers and digital infrastructures are significant consumers of energy, with ICT accounting for 5–9% of total electricity demand, projected to rise to 20% by 2030. Additionally, the proliferation of digital technologies contributes to a surge in electronic and electrical waste (e-waste), expected to reach 120 million tons annually by 2050. To tackle this challenge, circular economy principles must be applied to digital infrastructure, focusing not only on sustainability but also on ensuring a stable supply of critical raw materials. Digital technologies play a pivotal role in scaling up the circular economy by enabling new business models and complex supply chains. However, gaps in information and transparency hinder the full potential of the digital circular economy (DCE). The integration of enabling technologies such as functional electronics, blockchain, edge computing, UAVs, 5G/6G, big data, and AI is essential to enhance existing circular business approaches. The emerging concept of Industry 5.0, characterized by sustainable manufacturing and operator well-being, complements Industry 4.0 by prioritizing societal goals and environmental sustainability. While Industry 5.0 aims to improve smart factory efficiency through technology, it also emphasizes the minimization of environmental and social impacts, aligning with evolving societal trends. Despite the prevalence of cloud-based IoT applications, centralized solutions pose limitations, including vulnerability to attacks or maintenance issues. Decentralized alternatives are being explored to mitigate these risks and ensure system reliability. Artificial Intelligence (AI) enhances machine intelligence, enabling tasks such as recommendation systems and autonomous decision-making. In IoT systems, AI processes data collected by sensors and makes decisions, typically within centralized cloud architectures. Training AI systems to solve specific problems is crucial for their effectiveness in IoT applications.

9.16 Automation and Robotics in the Circular Economy of the Future (Nishal et al. 2023; Ozkan-Ozen et al. 2020)

Robots are a significant contributor to the reduction of material waste because of their capacity to perform exact actions frequently and reliably. When a result, the market for robots has taken off. According to the most recent study from the International Federation of Robots (IFR), 2021 will go down in history when the number of new industrial robots deployed in factories worldwide reaches an all-time high of over 500,000. Although employing robots clearly benefits users greatly, this rise is good for robot producers. However, what happens to robots when their operational lives are coming to an end? Manufacturers are always looking for methods to lessen their environmental impact, and a circular economy relies heavily on the responsible disposal of robots. Services that guarantee that their robots contribute to resolving global sustainability issues will aid in this endeavour. Numerous automotive factories integrate robotics components such as motors and pumps into their operations, often at rapid rates. These robots are tasked with various operations to maximize efficiency, including wheel assembly and windscreen installation, leveraging their exceptional accuracy and repeatability. Processes like trim removal and cuts, commonly carried out through methods like die casting, plastic molding, and material cutting, are also executed by robots due to their precision. Mobile autonomous robots, equipped with vision systems and proprietary algorithms, navigate factory and warehouse spaces, evolving from the historically used autonomously driven vehicles (AGV) in production facilities. However, the logistical landscape has faced challenges, exacerbated by several suppliers and dealers altering their policies regarding order delivery and receiving, thereby intensifying the need for automation in handling such tasks. Jobs that are done in-store are moved to a factory or delivery facility, where order fulfillment expenses go up. Robots rely on actuator systems, including DC engines, servo motors, steppers, etc., to perform motion and manipulation operations in the real world. The selected parts have to meet the architecture and operational requirements of the robots. Currently, robot designers size drives primarily by trial and error. The component size is raised repeatedly until the performance parameters are met. The actuators are sized using a single operational torque/speed point. Mobile robots with new characteristics that can improve cooperation have been outfitted with the newest sophisticated technologies. It allows positioning, locating, and navigation activities to be monitored by the client/user directly from any robot network computer.

9.17 Forward-Thinking Viewpoints on Robots in Sustainable Production and Recycling (Akbari 2024; de Mattos Nascimento et al. 2024; Kiran and Sahin 2024; Kumar et al. 2024b)

In the manufacturing industry, robots play a critical role in promoting a circular economy. They play a key role in the processes of remanufacturing and refurbishing outdated items to bring them back to functional state. Robots are able to precisely disassemble things and separate parts that are meant to be reused. By employing recycled materials to construct items, robots help create a production cycle that is less dependent on virgin resources and more sustainable. By doing this, the production process's carbon footprint is decreased and natural resources are preserved. Modern robots' data analytics skills help to optimise supply chains, guarantee the most sustainable sourcing and utilisation of commodities, and promote a manufacturing environment that aligns with the circular economy's tenets. Recycling has changed significantly with the introduction of recycling robots. At first, human labour was unable to manage all recyclable materials. Dismantling LCD displays and monitors was dangerous for human workers because of materials like mercury and other gases, as well as parts with sharp edges. This acted as a barrier to the development of a sustainable circular economy as not all of the priceless elements found in items such as LCDs could be extracted and recycled again. Companies took the wonderful idea to construct recycling robots that could undertake recycling jobs that were deemed detrimental to human workers in order to address this issue and reduce the associated hazards. The automated KUKA KR QUANTEC industrial robot, created by the Irish technology company Votechnik, is a prominent example of an industrial robot.

The KUKA KR QUANTEC eliminates hazardous components like fluorescent bulbs and displays as well as toxic fumes from a cell. These materials can be recycled safely by recycling robots. Thus, the circular economy's tenets are maintained and the resources' usage is assured. One of the major sources of pollution in the environment is LCD televisions. About 200 million TVs are sold each year. This suggests that massive waste streams are produced, and the safe, economical, and effective disposal of this garbage is necessary. Environmental investigation indicates that "hidden" greenhouse gases such nitrogen trifluoride are present in LCD gadgets and flat-screen TVs. The fact that these gases can hasten global warming is good enough justification to ban them. That isn't the case, though. Demand from consumers for these LCD screens is always rising, necessitating the production and usage of an increasing amount of these gases. As a result, both the manufacturing cost and the waste stream continue to rise. These toxic gases may be safely collected, repurposed, and reused when recycling robots manage the recycling of these equipment. Therefore, it is possible to make more LCD gadgets and flat-screen TVs without releasing harmful gases. This lowers production costs and allows for the continuing use of important resources. Robots that recycle are meant to reinforce and enhance a circular economy that is sustainable. The robots for recycling are designed to manage electrical trash.

Recycled electronic consumer items are essential to a sustainable circular economy. However, workers who recycle electronic garbage may sustain injury. Therefore, there has to be the best possible uses for automation and robots. The procedures involved in recycling need to be reduced to guarantee active involvement. This suggests that in order to guarantee recycling is successful, efficient, and clean, automation and intelligent robots must be employed. In this manner, more process enhancements are made. AI has the ability to differentiate between various materials using sophisticated visual systems. This does not imply that recycling will completely exclude human labour. Instead, human labourers and recycling robots can collaborate to recycle resources. Recycling processes involving hazardous materials and gases can be handled by robots. Human workers can also do tasks that call for human skill. In the recycling industry, there would be less exploitation and more safety in this way. Robotic recycling may result in fewer employment opportunities initially for human workers, but it will increase the amount of jobs in other sectors like engineering, maintenance, and oversight. In the recycling sector, robotics and artificial intelligence have established industry norms. At the Material Recovery Facilities, comprehensive and effective recovery of precious materials from waste streams is guaranteed by AI-driven robots.

9.18 Automation Breakthroughs for Circular Supply Chain Operations Are Anticipated (Junaid et al. 2024)

Manufacturing is the process of producing items using labour, equipment, procedures, and machinery. Recently, we have become more ambitious in the way we develop and create new items. Artificial intelligence and robots are two examples of new technology that may expand our thinking and help us create more creative, sustainable, and circular goods. However, precisely what are these technologies?. An interdisciplinary discipline that combines engineering and computer science is robotics. It involves the creation, assembly, programming, use, and maintenance of robots. A sub field of computer science called artificial intelligence (AI) focuses on creating computer programmes that can carry out activities that would typically need human intellect. Artificial intelligence (AI) algorithms are capable of learning, seeing, solving problems, comprehending language, and using logic. As you can see, artificial intelligence and robotics are actually two different technologies. While AI includes programming "intelligence," robotics focuses on creating and managing robots. They can, however, combine to form AI robots—robots that are guided by AI algorithms. The Fourth Industrial Revolution, or Industry 4.0, is the manufacturing future that these AI robots are leading.

9.19 Case Examples in the Future that Highlight Robotics' Potential for Circularity (Mastos et al. 2021; Veile et al. 2024)

The advancement of Industry 4.0 initiatives hinges on the integration of advanced robots and their synergy with machine vision, impacting every facet of industrial operations. Industrial robotics and machine vision have evolved in tandem, driven by the increased robustness and reliability of cameras in demanding production environments. This evolution is propelled by advancements in robotic technologies, establishing them as highly efficient and profitable methods for industrial tasks. Incorporating smart cameras into robotic arms enables a broad range of applications, allowing the arm to manipulate components for various inspections. Machine vision combines multiple techniques to capture and interpret images for robot-based inspection and guidance, facilitating surface fault detection, color analysis, presence/absence checks, inspection, and growth zone identification. Computer vision technologies provide essential components for developing efficient solutions in data collection, anomaly detection, quality control, and product testing, ultimately enhancing productivity and minimizing the need for human intervention. In industries such as car painting, which poses challenges due to toxicity and the shortage of skilled painters, robotic arms play a crucial role in ensuring durability and consistency of paint coatings. Robots follow predetermined paths to minimize waste and cover large areas efficiently, with additional applications in sealant and adhesive applications suited for mass production. Cobots, smaller collaborative robots, handle tasks like loading, unloading, and machine tending in smaller industrial settings, contributing to labor cost reduction through error-free repetitive tasks. Light robots equipped with force-sensing technology excel in dynamic operations, offering significant potential across various industrial applications.

9.20 Production

In the production process, this technology plays a vital part in carrying out all current operations. In the metals sector, electronic process management is used throughout production operations when hot metal ingots are rolled into final shapes like coils and strips. Although aluminium and other metals are currently manufactured in a manner similar to this, the steel industry was the first to employ it. Robots are used in a contemporary steel plant to do hot rolling under machine control. A big, hot metal billet is moved through a rolling mill, which consists of one or more sets of giant cylindrical rolls that compress the metal and lower its cross section, in order to complete the rolling process. It will take many passes to gradually decrease the ingot to the appropriate thickness. The ingot's temperature and readings are measured by automated equipment and sensors after each run through the rolls, and the roll settings are controlled by the control computer.

9.20.1 *Farming*

The agricultural sector has been actively pursuing the use of several robotic technology platforms in order to enhance efficiency and reduce overall expenses. Tractors and harvesters guided by GPS have also become common among farmers. Lately, there has been a surge in the observational use of self-reliant systems that perform tasks like weed repelling, watering, mowing, pruning, and thinning.

9.20.2 *Use in the Kitchen and Around the House*

A very expensive advancement in robotic kitchen technology is about to be released. It can be controlled by a smartphone, which enables the robot to choose a menu and arrange pre-packed ingredient containers before swiftly and effectively cooking the pre-planned meal. Robotics also develops a smart refrigerator and dishwasher that are affordable for consumers, complete with a smart kitchen robot.

9.20.3 *Robotics in Medical Procedures*

The development of robotics has the potential to revolutionise many aspects of healthcare, including surgery, recuperation, therapy, companionship for patients, and routine activities. Health care personnel are not meant to be replaced by robotic equipment; rather, it is meant to make their tasks easier. This technology is utilised to provide food, medication, and other necessities to affected patients during the COVID-19 epidemic.

9.20.4 *Matching Human Ability with Industrial Robotics*

This is becoming progressively more advanced, quicker, cheaper, and able to do more tasks, manufacturing is about to undergo a revolution. They compete with more human abilities and traits including experience, memory, trainability, and detection. They are thus in competition for more occupations, like packaging and picking. In the manufacturing sector, robots have long guaranteed performance, efficiency, and speed. Robots are revolutionizing production and assembly processes by carrying out a variety of specialized tasks and component transfers in manufacturing settings. They are particularly valuable for handling hazardous or repetitive tasks that may cause fatigue and injuries to human workers. By deploying robots in these roles, companies can improve workplace safety, enhance efficiency, and reduce the physical strain on human laborers.

9.20.5 The Automobile Sector

The automotive sector, recognized as one of the most advanced industries globally, has elevated automation standards by incorporating collaborative robotics into its assembly processes for the first time. Encouraging factory workers to accept interactive robotics is a common practice. Contemporary robots can help employees work more productively and outperform rivals. It is projected that the number of robots used in the car industry to produce automobiles will increase even more. As the pioneer of Industry 4.0, the automobile industry is introducing collaborative small-scale robotics on the mounting side. Previously, staff members were replaced by robots for solitary or hazardous tasks. Robots that aid humans are increasingly being used in the car production industry. One of the fastest-growing industries when it comes to the use of industrial robotic technology has always been the automobile sector. Almost every facet of the automobile manufacturing process involves robots, making it one of the world's most highly automated supply chains.

9.20.6 Storage and Transportation

Initially employed in die casting, robots have expanded their role in manufacturing, now fulfilling a wide array of tasks in logistics and warehousing across various industries. The proliferation of robotics in manufacturing can be attributed to advancements in software and sensor technologies. Some of the latest and most intriguing robotic applications are observed in factories, warehouses, and distribution centers. In warehouse environments, robots offer enhanced convenience with plug-and-play setups, enabling swift integration and interoperability among systems. Beyond ensuring multifaceted consistency and resilience in diverse assembly conditions, robots are capable of performing additional tasks as well.

9.21 Beyond Tomorrow: New Developments (Kumar et al. 2024b; Shakur et al. 2024)

9.21.1 Forecasts on Upcoming Developments and Their Possible Effects

The best phrase to create and discuss the social, environmental, and economic facets of sustainability is "sustainable manufacturing." SM has close ties to I4.0, one of whose pillars being the advancement of sustainability in organisations. I4.0's production flexibility and capacity to employ renewable energy sources make it a valuable tool for the advancement of sustainability. Additionally, I4.0 technologies can boost productivity through supply chain integration or increase resource efficiency.

Industrial managers now have the opportunity to surpass traditional environmental control measures and integrate intelligent process safety and control, along with prioritizing the well-being of both the community and employees. Consequently, Industry 4.0 not only serves as a strategy for enhancing production efficiency but also emphasizes social, environmental, and economic advancements as crucial priorities.—the cornerstones of sustainability, also referred to as the triple bottom line (TBL). However, current research indicates that a number of obstacles still need to be addressed before sustainable manufacturing can be achieved. These include the application of life cycle investigation, betterment of supply chain management and maintenance, mitigation of environmental impacts, adoption of I4.0 technology frameworks, appropriate application of circular economy concepts, real-time data processing, and security. Optimisation of the maintenance schedule, energy resource management, and other areas. To meet these issues, an organization's culture of innovation and learning is needed, along with the application of academic knowledge to practice in order to support the adoption of sustainable practices and developing technology. This kind of competence might be attained by introducing sustainable manufacturing ideas through innovative training techniques and enhancing comprehension of upcoming technologies and associated concepts through I4.0 skills training.

The inclusion of the socio environmental link, given that this is a factor that gives humans the ability to survive. In a world where pollution, biodiversity loss, and resource depletion are all major environmental issues, it is critical to implement more sustainable socio-technical systems. Furthermore, even though CE has not been fully integrated with I4.0 technologies, it has shown promise in addressing sustainability issues. Sustainability is defined as the capacity of the economy and environment to coexist, providing opportunities for businesses to profit while posing no risks to the environment. These days, the CE includes a revitalizing component and intent as a rhythm that examines the many stages of material and energy use, enabling resource investigation and observation. However, obstacles pertaining to cultural, technological, legal, and marketing elements still need to be removed before CE can be widely implemented. A further difficulty is that CE appears to be in its infancy and early phases of development, making it difficult to evaluate and monitor using the indicators and evaluation methods available. Extending research on closed-loop supply chain models is necessary to further efficient techniques of CE, since identifying metrics can improve existing methodologies. According to this viewpoint, further research on the CE and I4.0 is also required to create a plan of action that may facilitate social transformations and create a worldwide shift towards a sustainable circular economy by redesigning modes of production and consumption to more environmentally friendly ones. Though CE has been discussed, there is now some progress over how it integrates with I4.0 pillars and how innovation strategies for I4.0 deployment with an emphasis on sustainable development compare to each other. In order to embrace new ideas and approaches from various schools of thought, researchers should focus on finding connections between circular economics, the pursuit of shared economies, dematerialization, the Internet of Things, and production systems. This will allow for the adoption of innovative practices and sustainable systems of production that prioritise value and sustainable development.

9.22 Difficulties in Implementing I4.0 (Junaid et al. 2024; Narkhede et al. 2024)

The digitization of conventional operations and processes to increase productivity, transparency, and new business models is known as Industry 4.0. Since it's still in the early stages of deployment, there are several obstacles and hurdles. The following are a few of the main obstacles that professionals in the field encounter:

9.22.1 *Inadequate Adoption Research and Development*

Industry 4.0 and Smart Factory are relatively new concepts, hence there aren't any established models, regulations, or best practices for implementation. Large and small organisations alike may benefit from this transformation, which is being worked towards by a multitude of scholars and practitioners globally. There is no perfect universal solution, product, or model because every organisation is unique in its activities and structure.

9.22.2 *Legal Matters*

Industry 4.0 refers to a cyber-physical network in which a vast quantity of data is exchanged between sensors, machines, cloud computing, online applications, facilities, and people. To prevent concerns down the road, adequate legal compliance and issues should be taken into consideration prior to embracing digitalization. Adopting data-driven business models should be preceded by careful consideration of privacy and security safeguards. Project managers are required to perform appropriate legal research.

9.22.3 *The Company's Inadequate Digitization Vision and Strategy*

Industry 4.0 is a novel, cutting-edge, and intricate approach to company operations, thus before adopting it, organisations must have a very clear implementation vision and plan. There may be a lot of uncertainties on a regular basis, but the business, management, and staff should be prepared for them and have a plan in place to handle the situation. Workshops for proper staff training should be held so they may learn about the adoption process in its entirety as well as their new working style.

9.22.4 Indeterminate Financial Gains

Industry 4.0's massive data volume and supply chain's digitization present opportunities for developing new business models. However, many businesses were unaware of the many methods available to boost revenue and profitability prior to its adoption. Investigating various revenue streams is crucial, as is figuring out the ROI and payback duration.

9.22.5 Significant Funds and Investment Needed

Industry 4.0 calls for a massive investment in machines, cloud infrastructure, skilled labour, sensors, legal fees, software and hardware costs, RFID tags, operations, staff training programmes, backups, and other related items. Industry 4.0 is the complete conversion of traditional operations, processes, and systems to a digital model. Making a financial model of the complete system requires weighing the advantages and disadvantages of each and every expense.

9.22.6 Outdated IT Frameworks

Older IT infrastructure may not work with more recent software for production data analysis, forcing manufacturers to update their systems, which takes time.

9.22.7 Cooperation

Numerous factories operate with a blend of legacy and modern machinery, often lacking sensors and internet connectivity required for smart factory capabilities. However, manufacturers are hesitant to replace the older machinery due to the impracticality of retrofitting them with necessary upgrades. Additionally, a lot of firms lack IT systems that can analyse data that is received from devices that are linked.

9.22.8 *Skills Gap in the Workforce*

The implementation of Industry 4.0 poses human resources challenges for firms, particularly in attracting talent and retraining current staff due to many workers lacking expertise in integrating digital systems with industrial activities. Manufacturers often face the need to retrain employees on utilizing tablets, touch screens, and other gadgets to communicate with networked systems and optimize production processes using data-driven insights.

9.22.9 *Security Online*

In the past, manufacturing equipment was protected from cyber attacks by having as little of it as possible connected to the open internet. I4.0 adopts a different strategy by using the internet to link machines and corporate administration systems. Decision makers in the Industrial Internet of Things (IIoT) commonly raise query over the security of Industry 4.0 technologies.

9.22.10 *A Shift in Culture*

Effectively managing culture change is paramount to the success of Industry 4.0 initiatives, as poorly executed changes can have detrimental effects. Employees may exhibit hesitancy, resistance, or inability to adapt if they are not adequately prepared for the changes ahead. Therefore, gaining their support and preparing them for the technological transition is vital for the success of any digital endeavour. Leadership plays a significant role, particularly top-down leadership, in facilitating the necessary cultural shift for successful digital transformation.

9.23 Conclusion

The convergence of Industry 4.0 and the circular economy represents a paradigm shift in how industries approach supply chain management, production, and resource utilization. By leveraging the capabilities of I4.0 technologies, such as the Internet of Things, big data, robotics, and cloud computing, industries can unlock significant efficiency gains while aligning with sustainability objectives. One of the core contributions of I4.0 to the circular economy is the enhancement of transparency and real-time monitoring across supply chains. These technologies allow businesses to track resource use, monitor product life cycles, and manage waste more effectively, thus fostering the principles of reduce, reuse, and recycle at scale. Moreover, predictive maintenance enabled by I4.0 can significantly extend product life spans, reducing the need for raw material extraction and minimizing waste. Robotics and additive manufacturing further reduce energy consumption and material waste, creating more precise production processes that align with the goals of a circular economy. However, the transition to these advanced systems presents challenges. Companies must make substantial investments in infrastructure, address workforce skills gaps, and ensure cybersecurity in increasingly connected environments. Additionally, the shift requires a cultural transformation within organizations, promoting an innovative mindset that embraces digitalization and sustainability simultaneously. In conclusion, while the road to fully integrating I4.0 and circular economy principles is fraught with challenges, the potential benefits are immense. The future of sustainable supply chains hinges on this integration, offering businesses the opportunity to not only improve their operational efficiency but also significantly contribute to global sustainability goals. By adopting these emerging technologies, industries can move towards a future where economic growth is decoupled from resource consumption, leading to more resilient and environmentally conscious supply chains.

References

Akbari M (2024) Revolutionizing supply chain and circular economy with edge computing: systematic review, research themes and future directions. Manag Decis 62(9):2875–2899

Bag S et al (2024) Building smart product-service systems capabilities for circular supply chains in the Industry 4.0 era. Transp Res Part E Logist Transp Rev 188:103625

de Mattos Nascimento DL et al (2024) Interrelationships between circular economy and Industry 4.0: a research agenda for sustainable supply chains. Bus Strateg Environ 33(2):575–596

Faisal MN (2023) Role of Industry 4.0 in circular supply chain management: a mixed-method analysis. J Enterp Inf Manag 36(1):303–322

Gebhardt M et al (2022) Industry 4.0 technologies as enablers of collaboration in circular supply chains: a systematic literature review. Int J Prod Res 60(23):6967–6995

Junaid M et al (2024) Creating a sustainable future through Industry 4.0 technologies: untying the role of circular economy practices and supply chain visibility. Bus Strateg Environ 33:5753

Karmaker CL et al (2023) Impact of Industry 4.0 technologies on sustainable supply chain performance: the mediating role of green supply chain management practices and circular economy. J Clean Prod 419:138249

Kayikci Y, Gozacan-Chase N, Rejeb A (2024) Blockchain entrepreneurship roles for circular supply chain transition. Bus Strateg Environ 33(2):197–222

Kiran F, Sahin B (2024) Circular supply chain practices, big data implementation and firm performance: mediating role of digital technology. Electron J Infor Syst Dev Ctries 90(1):e12290

Kumar A, Mangla SK, Kumar P (2024a) Barriers for adoption of Industry 4.0 in sustainable food supply chain: a circular economy perspective. Int J Product Perform Manag 73(2):385–411

Kumar D et al (2024b) IoT-enabled coordination for recommerce circular supply chain in the Industry 4.0 era. Internet Things 26:101140

Liu L, Song W, Liu Y (2023) Leveraging digital capabilities toward a circular economy: reinforcing sustainable supply chain management with Industry 4.0 technologies. Comput Ind Eng 178:109113

Lu H, Zhao G, Liu S (2024) Integrating circular economy and Industry 4.0 for sustainable supply chain management: a dynamic capability view. Prod Plan Control 35(2):170–186

Mastos TD et al (2021) Introducing an application of an Industry 4.0 solution for circular supply chain management. J Clean Prod 300:126886

Narkhede G et al (2024) Significance of Industry 4.0 technologies in major work functions of manufacturing for sustainable development of small and medium-sized enterprises. Bus Strategy Dev 7(1):e325

Nishal M et al (2023) Significance of additive manufacturing amidst the pandemic. Mater Today Proc 72:2540–2546

Ozkan-Ozen YD, Kazancoglu Y, Mangla SK (2020) Synchronized barriers for circular supply chains in industry 3.5/industry 4.0 transition for sustainable resource management. Resour Conserv Recycl 161:104986

Shakur MS et al (2024) Exploring the challenges of Industry 4.0 adoption in the FMCG sector: implications for resilient supply chain in emerging economy. Logistics 8(1):27

Ülkü MA, Bookbinder JH, Yun NY (2024) Leveraging Industry 4.0 technologies for sustainable humanitarian supply chains: evidence from the extant literature. Sustain For 16(3):1321

Veile JW et al (2024) The transformation of supply chain collaboration and design through Industry 4.0. Int J Log Res Appl 27(6):986–1014

Chapter 10
Closing the Loop: Unlocking Sustainable Development Goals Through Circular Supply Chains

Abstract A fresh and persuasive viewpoint on the supply chain sustainability field is provided by Circular Supply Chain Management (CSCM), which incorporates the circular economy's tenets into supply chain management. An alternative to the prevalent linear economic paradigm of both manufacturing and consumption is the circular economy, or CE. Circular supply chain management is the process of integrating the CE into supply chain management. This chapter sheds light on how the CE supply chain and sustainable development objectives are related. CESC methods are now one of the approaches for achieving the SDGs. The most prominent and often spoken about SDG targets are SDG 2, SDG 8, SDG 9, SDG 12, SDG 14, and SDG 15. CESC can be used to achieve the SDGs, possibly with the help of industry 4.0 technologies like big data, IoT, and data analysis. These technologies can help materials move through the supply chain by enabling information traceability and transparency. Usage of CESC, which is flexible and relatively broad, enables both SMEs and major corporations to use it. To attain sustainability, CESC may have a significant influence on the manufacturing, agricultural, waste management, and natural resource sectors. Given the broad scope of CESC and its adaptability to other industries, industry research ought to concentrate on providing more detailed information about the possible sectors.

Keywords Circular supply chain management · Circular economy · SDG · Sustainability · Industry 4.0 · Supply chain

10.1 Introduction

10.1.1 Circular Supply Chains

Circular Supply Chains (CSCs) represent a paradigm shift in the way the supply chain is conceived and managed. At its core, circularity challenges the linear 'take, make, dispose' model by fostering an interconnected system where resources are conserved, reused, and continually cycled (Kayikci et al. 2022). Compared to earlier

J. Kandasamy et al., *Circular Supply Chain Management*, Circular Economy and Sustainability, https://doi.org/10.1007/978-3-031-81750-2_10

models, circular supply chains do less environmental damage. They provide businesses with a methodical approach to material collection and recycling, extending the useful life of the materials rather than letting them accumulate in landfills or deteriorate natural areas. If nothing changes in the way things are obtained, produced, distributed, used, reclaimed, and regenerated, the world may soon run out of numerous natural resources at the current rate of consumption. Circular Economy (CE) is becoming recognized as an approach to sustainable development that calls for a thorough overhaul of the entire framework governing human activity (Purvis et al. 2023). It covers the supply chain's manufacturing procedures as well as its consuming activities as shown in Fig. 10.1.

10.1.2 Sustainable Development Goals

With tremendous efforts to address the world's biggest challenges, the United Nations' ambitious initiative, the Sustainable Development Goals (SDGs) is found. Envisioned as a blueprint for a better and more sustainable future, a set of 17 interlinked objectives is comprised of the SDGs, each designed to address specific facets of global development (Nayal et al. 2022). From eradicating poverty and hunger to ensuring environmental sustainability and fostering peace, these goals serve as a universal call to action for governments, businesses, and communities worldwide. Sustainable development guarantees equitable and sustainable economic growth, enhanced social welfare, and preserved or enhanced environmental quality (Winston 2022). This is a long-term strategy that takes ecological, social, and economic interdependence into account. In order to create a more just and sustainable world, it seeks to advance harmony. Few examples of sustainable development are renewable energy, sustainable agriculture, green building, sustainable transportation, waste reduction and recycling, conservation of natural resources. This section explains the fundamental definition and principles that emphasize Circular Supply Chains, exploring how they diverge from conventional approaches and lay the groundwork for sustainable development.

10.2 Importance of Circularity in Sustainable Development

The urgency to embrace circularity extends beyond mere environmental concerns; it is a hub for achieving holistic sustainable development. The concept of Circular Supply Chains aligns seamlessly with the broader goals of sustainable development by addressing the interconnected challenges of environmental degradation, social inequality, and economic disparities (Lahane and Kant 2022; Cammarano et al. 2022). This section will underscore the critical importance of circularity within the context of sustainable development, setting the stage for a comprehensive

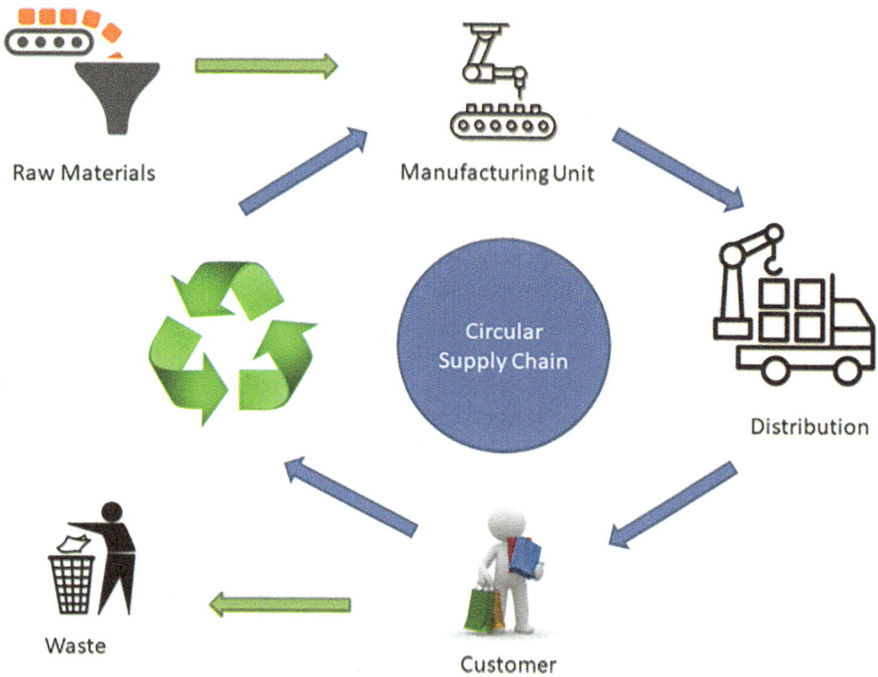

Fig. 10.1 Circular supply chains

exploration of how Circular Supply Chains become instrumental in achieving the ambitious targets set forth by the SDGs.

As the intricacies of Circular Supply Chains are navigated, their profound impact on specific SDGs will be unravelled, showcasing how sustainable practices can be pivoted towards by businesses and organizations, contributing meaningfully to a future where economic growth coexists harmoniously with environmental and social responsibility (Bag and Rahman 2023). A new paradigm for production and consumption known as the "circular economy" promises long-term, sustainable growth (Kandpal et al. 2024). By recycling garbage or giving it another chance as a fresh product, the circular economy allows us to recover waste, optimize resource use, and use fewer raw materials. A good correlation was observed between the Circular Economy and Sustainable Development Goal 12, which aims to attain a noteworthy decrease in waste generation by means of prevention, reduction, recycling, reuse, and the sustainable utilization of natural resources (Olabi et al. 2023). The journey towards achieving SDGs through Circular Supply Chains beckons us to embrace innovation, collaboration, and a commitment to a circular ethos that resonates with the sustainability imperatives of our time (El Wali et al. 2021).

10.3 Closing the Loop for Sustainable Development

In the complex global production scenario, the traditional linear approach of take-make-dispose is unsustainable and impacts our planet (Paliwal 2022). Every organization is looking for economic progress, and environmental gains and the focus on transitioning towards more sustainable practices is imperative. Businesses worldwide are looking into how to switch to CE business models. For the CE to become widely accepted, the UN system along with other multilateral bodies must endorse it. The circular economy also presents the possibility of far more effective action being taken to address the triple planetary catastrophe of pollution, climate change, and biodiversity loss (Rashid and Malik 2023). This is the exact kind of stimulant that the UN's failing 2030 Agenda for Sustainable Development should employ. The following section delves into the transformative scope of Circular Supply Chains (CSC) and explains how their adoption serves as a powerful medium for realizing the United Nations' Sustainable Development Goals (SDGs).

10.4 Linking SDGs to Global Challenges and the Need for Sustainable Solutions

The beginning of the Sustainable Development Goals is found in the acknowledgment of the complex challenges confronting humanity in the twenty-first century. From climate change and social inequality to poverty and environmental degradation, a comprehensive framework is provided by the SDGs to tackle these interconnected issues. The intrinsic link between the SDGs and the formidable global challenges industries face will be elucidated in this subsection, emphasizing the imperative for sustainable solutions (Yakovleva et al. 2019). By examining the root causes of these challenges and understanding their far-reaching implications, the way is paved for exploring how Circular Supply Chains (CSC) can emerge as a pivotal strategy in achieving these ambitious goals (Khalili-Fard et al. 2024).

As this exploration is embarked upon, it becomes evident that the SDGs are not mere aspirations but a shared commitment to building a more equitable, resilient, and sustainable world (Ogunmakinde et al. 2022). The subsequent sections of this chapter will delve deeper into the symbiotic relationship between Circular Supply Chains and specific SDGs, illustrating how the principles of circularity align seamlessly with the broader vision of sustainable development outlined by the United Nations.

10.5 Identification of Common Goals Between Circular Supply Chain Management (CSCM) and SDGs

At the intersection of Circular Supply Chain Management (CSCM) and the Sustainable Development Goals (SDGs), common goals are identified. The alignment between these two frameworks is discerned through an examination of shared objectives that collectively contribute to the advancement of sustainable development. This subsection explores how the Circular Supply Chain, and the SDGs are interconnected, highlighting the synergy that emerges when sustainable practices are integrated into supply chain processes. In the connection where Circular Supply Chain Management (CSCM) intersects with the Sustainable Development Goals (SDGs), a passive process unfolds, revealing shared objectives that support the alignment of these two key frameworks. Common goals are determined through a thorough exploration of their inherent principles, which collectively contribute to the progression of sustainable development. The alignment is not merely a consequence of conscious efforts but is rooted in their fundamental principles. This passive identification becomes a cornerstone for the subsequent exploration of their collective impact on fostering sustainability within global supply chain processes (Valverde and Avilés-Palacios 2021).

There needs to be a comprehensive understanding of how the passive identification of common goals serves as the foundation for a collaborative and symbiotic relationship. Further in the chapter, a deeper exploration awaits, shedding light on how these passive identifications actively shape sustainable practices, contributing meaningfully to the realization of specific SDGs. This embarks on a journey of discovery, illuminating the interconnectedness of CSCM and the SDGs. Each facet of CSCM, from resource efficiency to closed-loop systems, resonates with the aspirations enshrined within the SDGs (Kazancoglu et al. 2021). The meticulous reduction of waste through circular practices echoes the call for responsible consumption and production (SDG 12). The extension of product lifecycles aligns with the pursuit of sustainable infrastructure and innovation (SDG 9). The very essence of circularity—the decoupling of economic growth from resource depletion—finds harmonious alignment with the ambitious goal of climate action (SDG 13).

The synergy becomes evident when sustainable practices are integrated into supply chain processes becomes undeniable (Singh 2024). By embracing circularity, businesses become instrumental in achieving the aspirations outlined in the SDGs. Pollution prevention initiatives embedded within circular models contribute directly to clean water and sanitation (SDG 6). Responsible sourcing practices resonate with the quest for decent work and economic growth (SDG 8). The very act of closing the loop through material recovery aligns seamlessly with the pursuit of life on land (SDG 15).

10.6 Circular Supply Chain Management Practices: Examples, Outcomes, and Contributions to SDGs

In response to the growing imperative of sustainability, businesses are increasingly turning to Circular Supply Chain Management (CSCM) as a strategic framework to minimize environmental impact, enhance resource efficiency, and contribute to broader global sustainability goals (Montag 2023). Some of the key areas in which CSCM can contribute are waste reduction, clean energy and sustainable and resilient economy (Tseng et al. 2023). This section explores key CSCM practices through real-world examples, delving into their outcomes and elucidating the direct contributions these practices make towards achieving the United Nations' Sustainable Development Goals (SDGs). Table 10.1 illustrates the intersection of CSCM practices and their outcomes with SDG goals.

In essence, CSCM and the SDGs orchestrate a symphony of sustainable ambition. Their intertwined melodies offer a powerful counterpoint to the discordant harmonies of the "take-make-waste" model. By aligning business practices with the global goals for sustainable development, CSCM emerges as a powerful conductor, guiding us toward a future where environmental responsibility and economic prosperity resonate in perfect harmony (Shrimali and Agrawal 2024).

Table 10.1 Intersection of CSCM and SDG

Practice	Outcomes	SDG
Design for circularity	Resource conservation	Goal 12 (Responsible Consumption and Production).
	Extended product lifespan	Goal 13: Climate Action: Resource efficiency and reduced emissions contribute to climate action.
Reverse logistics	Reverse logistics	Goal 12: Emphasizes responsible consumption and production.
	Promotion of circular economy	Goal 8: Decent Work and Economic Growth: Circular practices create job opportunities and foster economic growth.
Waste reduction and recycling	Resource Conservation: Reclaimed materials are used to manufacture new products	Goal 12: Closed-loop recycling system minimizes waste and conserves resources.
	Reduced Environmental Impact: Closed-loop recycling minimizes the environmental impact	Goal 9: Sustainable infrastructure and innovation
Sustainable sourcing	Ethical Sourcing: Contribute to ethical practices by sourcing materials responsibly.	Goal 12: This contributes to a circular supply chain by promoting responsible sourcing and aligning.
	Environmental Stewardship: Sustainable sourcing practices reduce the environmental impact.	Goal 13: Climate Action: Resource efficiency and reduced emissions contribute to climate action.

10.6.1 Design for Circularity

One cornerstone of Circular Supply Chain Management is the incorporation of circular design principles. Companies are rethinking product design to prioritize longevity, reparability, and the ability to reclaim and reuse materials (Roskladka et al. 2023). Reuse, recycling, and material reduction are encouraged across your value chain when circularity and waste reduction are designed for. Reusing or recycling materials can lower the number of emissions in your value chain since it eliminates the need to extract and process new resources. Reduced costs, more efficient use of resources, and a sustainable business model are all benefits of circularity and waste reduction strategies. Circular design minimizes the need for new raw materials, conserving natural resources. Products designed for circularity have longer lifespans, reducing the frequency of replacements and associated environmental impacts.

Reduction of waste and circularity can also have important consequences for environmental justice. These methods can lessen your company's vulnerability to adverse effects on certain communities and present an opportunity for your business to think more deliberately about its waste, even while they do not ensure favourable environmental justice outcomes. Actions like designing sustainable and safer products, its packaging methods and developing circular solutions will help to design the circularity.

Example: Modular Smartphones
A notable example is the concept of modular smartphones. By allowing users to upgrade individual components such as cameras or batteries, rather than replacing the entire device, manufacturers extend the product's lifespan and significantly reduce electronic waste. This exemplifies how design for circularity actively contributes to SDG 12 (Responsible Consumption and Production).

10.6.2 Reverse Logistics

Reverse logistics, which entails gathering goods and commodities after they have been used or distributed, is an essential activity in supply chain management. This practice concentrates on a few main areas like remanufacturing, logistics management for unsold goods and undelivered products. Efficient management of product returns and end-of-life products is a hallmark of CSCM. Companies embracing reverse logistics effectively collect, refurbish, and recycle products, contributing to a closed-loop system (Ding et al. 2023). Reverse logistics' primary goal is to optimize the flow of goods, resources, and materials from the final customer to the supply chain's origin while reducing the environmental effect and maximizing its value. The key concept in reverse logistics is reduce, reuse and recycle. Reverse logistics' primary goal is to optimize the flow of goods, resources, and materials from the final customer to the supply chain's origin while reducing the environmental effect and maximizing its value.

The processes involved in reverse logistics are collecting the goods, sorting the goods, return to the warehouse, destruction of irreplaceable goods, recycling and recovering the materials. By encouraging sustainable production and consumption, reverse logistics is a supply chain technique that can aid in the achievement of the Sustainable Development Goal (SDG 12).

Example: Apple
Apple has been one of the profitable enterprises which employs Reverse logistics operations successfully. Apple has adopted "Reverse Logistics" to improve the environmental impact of their production process and win over customers. Apple produces a wide range of goods, including the iPhone, MacBook, and other items, which it sells in its retail locations worldwide. When a consumer wants to upgrade, Apple allows them to return their old phone. Customers give them their old phones in exchange for a discount on new phones, which are then taken to the company's factories where the parts are used to make new phones.

In this approach, they produce goods in an environmentally beneficial manner in addition to making more profit.

10.6.3 Waste Reduction and Recycling

Minimizing waste generation is a fundamental aspect of Circular Supply Chain Management. Reducing or decreasing the quantity of materials utilized to produce garbage is known as waste reduction. Companies adopting waste reduction and recycling practices focus on optimizing production processes to minimize environmental impact (Attiq et al. 2021). Key methods to reduce waste are purchase in bulk, use reusable items, buy second-hand or refurbished products, compost food wastes and leaves, repair and donate. The 7Rs of waste reduction are Rethink, Refuse, Reduce, Reuse, Repair, Regift and Recycle.

Recycling is the act of gathering, processing, and creating new goods out of resources that would have been thrown away as waste. Recycling is good for the environment, the economy, and your community. Goods ought to be recycled only when they are no longer able to be reused. Several SDGs like SDG 3 (Good health and well-being), SDG 8 (Decent work and economic growth) and SDG 12 can benefit from waste reduction and recycling.

Example: Subaru
Subaru recycles or utilizes everything again. Indeed, for more than a decade, Subaru's two manufacturing facilities in Japan and its lone American facility have refrained from disposing of their garbage in nearby landfills. Subaru's dedication to zero waste has significantly impacted their financial results, even if it is excellent for the surroundings and the Subaru community. Subaru saw savings of $1 to $2 million a year after their U.S. plant achieved zero waste. Subaru reuses everything given by

their employees and about 95% of the vehicle components could be either be recycled or reused. Together with its dedication to become a 100% zero waste business, Subaru also shares its effective zero waste process with other businesses and organizations globally. This closed-loop recycling system minimizes waste and conserves resources, aligning with SDG 12.

10.6.4 Sustainable Sourcing

It is a CSCM that extends beyond product design and end-of-life that considers to encompass sustainable sourcing practices. Companies emphasize responsible sourcing to ensure that raw materials meet ethical and environmental standards. Sustainable sourcing is an enterprise strategy that includes the selection of suppliers and the procurement of goods and services with consideration for social, economic, and environmental concerns. Reducing a business's detrimental effects on the natural world and society is the aim, as is fostering enduring connections with suppliers. Developing solid, long-lasting relationships with suppliers is a final objective of sustainable sourcing. Building these kinds of connections requires enhanced effectiveness on environmental, social, and ethical issues. A company's dedication to corporate responsibility is now reflected in its efforts to achieve this, which makes them an essential part of the entire organizational structure and business strategy. Companies can develop and strengthen their competitive advantages through efficient supply chain management, particularly in industries where production is mostly outsourced, like food and apparel.

Example: H&M

H&M, a clothing store, is dedicated to openness in the supply chain. They provide a database of 98.5% of their vendors' identities and locations on their web page and update it on every quarter. Anyone can check to see if their suppliers meet the requirements established by the business. This ensures that the behavior of their suppliers can be held openly responsible for them. H&M aims to use only 100% upcycled or environmentally friendly products by 2030 as part of its responsible sourcing commitment, in addition to disclosing information about its supply chain. As a result, H&M is a leader in ethical fashion retail and a trendsetter. Many people see issues with the fashion industry when it comes to ethical and ecological sourcing. This reputation could be altered if a well-known clothes shop like H&M can show that it is committed to sustainability and ethical business practices without sacrificing revenues. Other retailers might then take note and emulate this approach. This contributes to a circular supply chain by promoting responsible sourcing, aligning with SDG 12.

10.7 Future Trends and Opportunities in Circular Supply Chain Management

As Circular Supply Chain Management (CSCM) continues to evolve, it is essential to look ahead and anticipate future trends that will shape the landscape of sustainable supply chains. Recent trends in technological advancements in AI, blockchain and IIoT are likely to play a significant role in the evolution of CSCM and its potential impact on achieving Sustainable Development Goals (SDGs).

10.7.1 Artificial Intelligence

More people than ever adore AI. Our 2023 State of Production Report indicated that 85% of organizations have utilized AI technologies and foresee supply chain management improvements. Automation and AI may increase the efficiency of the supply chain and its competitiveness. In a survey, 97% of respondents said they thought AI would have an impact on product creation. AI for supply chain management can be used for Smart sourcing, inventories, monitoring supply chain and logistics.

10.7.2 Increased NPI Speed Focus

New product introduction (NPI) is hampered by geopolitical unrest, inflation, and economic instability. 94% of manufacturers face NPI obstacles, and 49% desire to accelerate NPI, according to our most recent State of Manufacturing Report. In order to maintain rapid prototyping and innovation, organizations must transition to more robust and efficient supply chains in order to meet the need for faster NPI. Working with a supplier who uses AI ensures a robust supply chain, quick prototyping, and a shorter time to market. Detecting material selection and supply chain issues can also be aided by early communication between the acquisition and distribution chain management teams.

10.7.3 Customisation

Working with a supplier who uses AI ensures a robust supply chain, quick prototyping, and a shorter time to market. Detecting material selection and supply chain issues can also be aided by early communication between the acquisition and distribution chain management teams.

10.7.4 The Internet of Things

The term "Internet of Things" (IoT) describes a network of physical items that are wirelessly networked, digitally connected, and easily accessible from any location. By 2030, there will be over 25.4 billion IoT devices in use, according to research done by Dataprot. For even more advantages, IoT devices can be combined with other technological components. For instance, the information generated by a sensor in your storage facilities might offer useful data to assist in automating other activities, including as prediction and asset tracking.

10.7.5 Digitalization

According to a recent study, 31% of respondents are using digital supply chain analytics and visual tools, and 60% of respondents now use them. Digitalization of the supply chain has many advantages like decreasing supply chain disruptions, automated workflows and real-time product tracking.

10.7.6 Better Relationships in SCM

Relationships may be made more cooperative and productive by putting them at the core of supply chain management. When it comes to changing laws and regulations, buyers that have good working relationships with their vendors are in a better position to obtain their compliance. There are several distinct marketplaces in which strategic alliances might arise, as well as various combinations of providers and customers. The configurations of alliances can differ greatly. One supplier and one customer are involved in a typical supplier-customer relationship. In an industry, alliances can also form between two horizontal providers; one example of this is the partnership between Microsoft and Dell. Several parties, such as shipping companies that need to coordinate their efforts for multifaceted shipments, may be involved in a vertical supplier-supplier alliance.

10.7.7 Resilience and Risk Management

In a circular supply chain, resilience and risk management are critical because they enable supply chains to adjust and continue operating even in the face of disruptions. Manufacturers who embrace the circular economy can help advance environmental sustainability. Supply chains may adjust and continue to run in the event of unanticipated shocks by turning vulnerabilities into strengths, which promotes resilient and sustainable growth in a globalized society.

10.7.8 Enhanced Visibility

The capacity to monitor and comprehend each stage of a good or service's route from supplier to customer is known as supply chain visibility, or SCV. It entails gathering and evaluating data from every stage in the supply chain, especially ingredients, manufacturing, and logistics. Supply chain management technology, which incorporates IoT, 5G, AI, machine learning, robotics, and PLM (Product Lifecycle Management) software, is one of the most widely used tools for supply chain visibility.

10.8 Identifying Emerging Opportunities for Sustainable Development

While anticipating future trends is crucial, identifying emerging opportunities is equally vital. Businesses and organizations can leverage these opportunities to contribute meaningfully to sustainable development through circular supply chains.

10.8.1 Circular Business Models

Many industries can profit from circular models, but the built environment is one of the main ones. For starters, just 40% of construction debris is currently recycled or reused as buildings attain the end of their useful lives. This indicates that 60% has been squandered, which is expensive for the environment and the business. But while maintaining the use of materials and buildings, a circular business model that emphasizes recycling, reusing, and repairing can assist to reduce or even completely eliminate—waste and emissions from the industry. Retrofitting is one instance of this in architecture that helps prevent demolition and the requirement for more basic supplies to replace old buildings.

10.8.2 Eco-Innovations and Sustainable Sourcing

Eco-innovation is a novel approach to business that enhances a company's competitiveness and performance while simultaneously promoting sustainability over a product's whole life cycle. In addition to helping SMEs remain abreast of rules and standards, particularly those pertaining to the environment, it can also assist SMEs access new and wider markets, boost productivity, draw in new investment, and raise profitability throughout the value chain. Eco-innovations like create synergies, improve environmental performance, societal performance and communication can act as a pillar for sustainable development.

10.8.3 Circular Economy Regulations and Incentives

The existing linear economic paradigm creates barriers, which can be reduced in large part through the use of economic incentives. By pricing adverse effects, directing the market toward more circular approaches, and influencing consumer behavior to make more circular decisions, these incentives can alleviate market failures. The CE incentives include public procurement, green procurement, circular product design, circular financing, circular innovation and extended producer responsibility.

10.8.4 Data-Driven Decision-Making

Organizations may react swiftly to shifts in consumer preferences, market conditions, and competitive challenges by using data-driven decision-making. By constantly observing key indicators of success and market signals, businesses can quickly modify their supply chain strategy to remain ahead of the curve. In circular supply chains, data-driven analysis can be used to spot trends and possibilities. When deployed properly, a data-driven supply chain management system enables managers and operators to collaborate proactively and confidently to solve a wide range of sourcing, manufacturing, and delivery issues at every stage of the production process. A proactive and self-assured supply chain supported by robust data-driven analysis is the organization's best chance to raise sales, improve profit margins, and build brand loyalty and customer retention.

10.9 Bridging Future Trends and SDGs

10.9.1 Goal 6—Clean Water and Sanitation

A circular supply chain can help organizations contribute to the SDG 6, Water and Sanitation for All, by decreasing waste and increasing the use of materials, water, and energy.

10.9.2 Goal 8—Decent Work and Economic Growth

A circular supply chain can assist in achieving SDG 8 reducing waste via the reuse of resources and goods. Circular practices create job opportunities and foster economic growth.

10.9.3 Goal 9—Industry, Innovation, and Infrastructure

By reducing pollutants and waste through innovation in product design and production, circular supply chains contribute to the achievement of SDG 9, which is focused on industry, innovation, and infrastructure.

10.9.4 Goal 11—Sustainable Cities and Communities

Circular supply chains minimize waste and pollution, keep products in use, and regenerate natural systems, contributing to sustainable cities by improving resource efficiency, reducing environmental impact, and enhancing urban resilience.

10.9.5 Goal 12—Responsible Consumption and Production

CSCM promotes responsible consumption and sustainable production. Future trends promoting circular design and business models directly contribute to responsible consumption and production.

10.9.6 Goal 13—Climate Action

SDG 13 is supported by circular supply chains by facilitating the switch from petroleum and natural gas to renewable energy. Industries concentrating on resource efficiency and reduced emissions contribute to climate action.

10.9.7 Goal 14—Life below Water

SDG 14 is enhanced by a circular supply chain by decreasing waste and enhancing the sustainability of ocean resources. Other ways to improve is by protecting and rehabilitating water-related ecosystems, enhancing the quality of water by cutting down on pollution and getting rid of waste, increasing the efficiency of water use in all areas and using integrated water supply management into practice.

10.9.8 Goal 15—Life on Land

SDG 15, which is concerned with life on land, can be improved via circular supply chain methods by lowering resource waste, fostering sustainable development, green logistics and circular farming.

10.10 Conclusion

In conclusion, the trajectory towards sustainability is traced through CSCM, where each circular practice is transformed into a stepping stone to a more responsible, efficient, and resilient supply chain ecosystem. The transformative potential of CSCM is seen to extend far beyond individual businesses as it is recognized as a collective force that steers us toward a sustainable future. By embracing CSCM, businesses not only fortify their supply chains but actively contribute to a global narrative of sustainable growth, where economic prosperity converges with environmental and social responsibility. The role played by CSCM in shaping a future where businesses thrive as stewards of sustainability is highlighted, leaving an enduring legacy for generations to come.

Circular Supply Chain Management practices, exemplified through real-world cases, showcase not only positive outcomes for businesses but also active contributions toward achieving the SDGs. As organizations increasingly integrate circularity into their supply chains, the collective impact on sustainable development becomes more tangible, emphasizing the pivotal role CSCM plays in shaping a more sustainable and interconnected global future. The achievement of specific SDGs is facilitated by the contribution of circular supply chains. From environmental conservation to social responsibility, the ways in which circular supply chains actively contribute to the realization of specific SDGs are explored. The convergence of future trends in Circular Supply Chain Management with Sustainable Development Goals provides a unique opportunity for businesses and organizations to contribute to global sustainability efforts. This forward-thinking approach ensures that Circular Supply Chain Management remains a dynamic force for positive change in the global pursuit of sustainability.

References

Attiq S, Habib MD, Kaur P, Hasni MJS, Dhir A (2021) Drivers of food waste reduction behaviour in the household context. Food Qual Prefer 94:104300. https://doi.org/10.1016/j.foodqual.2021.104300

Bag S, Rahman MS (2023) The role of capabilities in shaping sustainable supply chain flexibility and enhancing circular economy-target performance: an empirical study. Supply Chain Manag Int J 28(1):162–178. https://doi.org/10.1108/SCM-05-2021-0246

Cammarano A, Perano M, Michelino F, Del Regno C, Caputo M (2022) SDG-oriented supply chains: business practices for procurement and distribution. Sustain For 14(3):1325. https://doi.org/10.3390/su14031325

Ding L, Wang T, Chan PW (2023) Forward and reverse logistics for circular economy in construction: a systematic literature review. J Clean Prod 388:135981. https://doi.org/10.1016/j.jclepro.2023.135981

El Wali M, Golroudbary SR, Kraslawski A (2021) Circular economy for phosphorus supply chain and its impact on social sustainable development goals. Sci Total Environ 777:146060. https://doi.org/10.1016/j.jclepro.2023.135981

Kandpal V, Jaswal A, Santibanez Gonzalez ED, Agarwal N (2024) Circular economy principles: shifting towards sustainable prosperity. In: Sustainable energy transition: circular economy and sustainable financing for environmental, social and governance (ESG) practices. Springer Nature, Cham, pp 125–165. https://doi.org/10.1007/978-3-031-52943-6_4

Kayikci Y, Kazancoglu Y, Gozacan-Chase N, Lafci C (2022) Analyzing the drivers of smart sustainable circular supply chain for sustainable development goals through stakeholder theory. Bus Strateg Environ 31(7):3335–3353. https://doi.org/10.1002/bse.3087

Kazancoglu I, Sagnak M, Kumar Mangla S, Kazancoglu Y (2021) Circular economy and the policy: a framework for improving the corporate environmental management in supply chains. Bus Strateg Environ 30(1):590–608. https://doi.org/10.1002/bse.2641

Khalili-Fard A, Parsaee S, Bakhshi A, Yazdani M, Aghsami A, Rabbani M (2024) Multi-objective optimization of closed-loop supply chains to achieve sustainable development goals in uncertain environments. Eng Appl Artif Intell 133:108052. https://doi.org/10.1016/j.engappai.2024.108052

Lahane S, Kant R (2022) Investigating the sustainable development goals derived due to adoption of circular economy practices. Waste Manag 143:1–14. https://doi.org/10.1016/j.wasman.2022.02.016

Montag L (2023) Circular economy and supply chains: definitions, conceptualizations, and research agenda of the circular supply chain framework. Circ Econ Sustain 3(1):35–75. https://doi.org/10.1007/s43615-022-00172-y

Nayal K, Kumar S, Raut RD, Queiroz MM, Priyadarshinee P, Narkhede BE (2022) Supply chain firm performance in circular economy and digital era to achieve sustainable development goals. Bus Strateg Environ 31(3):1058–1073

Ogunmakinde OE, Egbelakin T, Sher W (2022) Contributions of the circular economy to the UN sustainable development goals through sustainable construction. Resour Conserv Recycl 178:106023

Olabi AG, Shehata N, Sayed ET, Rodriguez C, Anyanwu RC, Russell C, Abdelkareem MA (2023) Role of microalgae in achieving sustainable development goals and circular economy. Sci Total Environ 854:158689. https://doi.org/10.1016/j.scitotenv.2022.158689

Paliwal R (2022) From waste to wealth: stepping toward sustainability through circular economy. In: Handbook of solid waste management: sustainability through circular economy. Springer Nature, Singapore, pp 441–463. https://doi.org/10.1007/978-981-16-4230-2_82

Purvis B, Celebi D, Pansera M (2023) A framework for a responsible circular economy. J Clean Prod 400:136679. https://doi.org/10.1016/j.jclepro.2023.136679

Rashid S, Malik SH (2023) Transition from a linear to a circular economy. In: Renewable energy in circular economy. Springer International Publishing, Cham, pp 1–20. https://doi.org/10.1007/978-3-031-42220-1_1

Roskladka N, Jaegler A, Miragliotta G (2023) From "right to repair" to "willingness to repair": exploring consumer's perspective to product lifecycle extension. J Clean Prod 432:139705. https://doi.org/10.1016/j.jclepro.2023.139705

Shrimali R, Agrawal NK (2024) Climate responsive insulation strategies: a comparative analysis for enhanced energy conservation and reduced environmental footprint in Indian urban contexts. Environ Dev Sustain:1–58. https://doi.org/10.1007/s10668-024-05267-0

Singh RK (2024) Wheeling towards sustainability: the nexus of external pressures, green innovation and circular supply chain excellence. Bus Process Manag J 30(4):1044–1064. https://doi.org/10.1108/BPMJ-10-2023-0772

Tseng ML, Li SX, Lim MK, Bui TD, Yuliyanto MR, Iranmanesh M (2023) Causality of circular supply chain management in small and medium-sized enterprises using qualitative information: a waste management practices approach in Indonesia. Ann Oper Res:1–42. https://doi.org/10.1007/s10479-023-05392-5

Valverde JM, Avilés-Palacios C (2021) Circular economy as a catalyst for progress towards the sustainable development goals: a positive relationship between two self-sufficient variables. Sustain For 13(22):12652

Winston N (2022) Sustainable community development: integrating social and environmental sustainability for sustainable housing and communities. Sustain Dev 30(1):191–202. https://doi.org/10.1002/sd.2238

Yakovleva N, Frei R, Rama Murthy S (2019) Editorial introduction: achieving sustainable development goals through sustainable supply chains in the post-global economy. Springer International Publishing, pp 1–9

Chapter 11
Investigating Circular Supply Chain Strategies and Industry 4.0 Implementation: A Case Analysis of a Paper Manufacturing Company

Abstract Environmentalists, ecologists, and governments have become interested in sustainability challenges from the perspectives of the economy, environment, and society. In this regard, the circular economy offers the chance to reduce production and consumption activities, reduce carbon footprints, and simultaneously convert waste into resources. Models and solutions for the circular economy supported by industry 4.0 technologies were created to create new products with diverse uses from ones that are nearing the final phase of their life cycle. In this chapter case study was done in the uses of Industry 4.0 technologies on a paper manufacturing industry. Using the framework that has been built, the research utilizes the Multi-Grade Fuzzy Approach (MGFA) to evaluate each selected organization's present level of Industry 4.0 and circularity readiness. A roadmap is also developed for the adaptation of circular strategies for paper manufacturing companies, industrialists and academicians to improve circularity in their supply chain.

Keywords Sustainability · Circular economy · Industry 4.0 · Multi-grade fuzzy approach · Supply chain · Life cycle

11.1 Introduction

In the ever-evolving landscape of supply chain management, the adoption of circular practices has gained significant attention for its potential to reconcile economic growth with environmental sustainability (Dey et al. 2022). This chapter delves into a compelling case study focusing on Industry 4.0 and Circular Supply Chain Management (CSCM) within the context of a prominent paper manufacturing organization located in South India—Case Organization. Further, this chapter embarks on a compelling journey into the heart of this paradigm shift, unravelling the intricacies of organizational readiness for Industry 4.0 and Circularity transformation. The foundation of this research rests upon an exhaustive examination of existing literature encompassing supply chain dynamics, Industry 4.0 tools, and Circularity practices. The practical applicability of the framework unfolds through the selection of

J. Kandasamy et al., *Circular Supply Chain Management*, Circular Economy and Sustainability, https://doi.org/10.1007/978-3-031-81750-2_11

case organization. The case organization was selected for supply chain maturity and digital adoption willingness. The Multi-Grade Fuzzy Approach (MGFA) is employed as a methodological lens, allowing for a nuanced assessment of each organization's current state of readiness.

With the insights gained from MGFA assessments, a toolkit is proposed for each case organization. These toolkits represent a synthesis of Industry 4.0 tools and Circular Supply Chain Management practices tailored to the unique needs and challenges of each sector. To ensure the robustness and applicability of the proposed toolkits, rigorous statistical validation is undertaken. This not only substantiates the feasibility of deploying Industry 4.0 tools and circular practices but also reinforces the research's practical contributions. The final phase of the research involves a conclusive assessment and validation with subject matter experts, culminating in the proposition of a comprehensive roadmap for Industry 4.0 and Circularity transformation. This roadmap synthesizes the lessons learned from diverse industries, offering actionable insights for organizations navigating the transformative journey toward a technologically advanced and circular future.

11.1.1 Development of a Framework for Digital Transformation

The literature study suggests that combining Industry 4.0 and Circularity transformation was not focused on earlier and thus it is considered as a gap. Based on the literature, a new framework for assessing Industry 4.0 and Circularity transformation is proposed as illustrated in Fig. 11.1. The conceptual model has been formulated from five perspectives of CSCM namely business, technology, sustainable development, collaboration, and management strategy to meet the requirements of the fourth industrial revolution.

The framework has three levels as tabulated in Table 11.1. The first level contains 5 enablers, which influence the industry 4.0 and Circularity transformation; the second level comprises 18 criteria, and the third level of 62 attributes.

11.2 About Case Organization—Paper Manufacturing Industry

The study was performed in a paper manufacturing organization located in South India. The paper organization manufactures writing paper and newsprint. The organization is an ISO 9001:2000 organization that distinguishes itself as a quality organization and is keen on implementing sustainable supply chain practices.

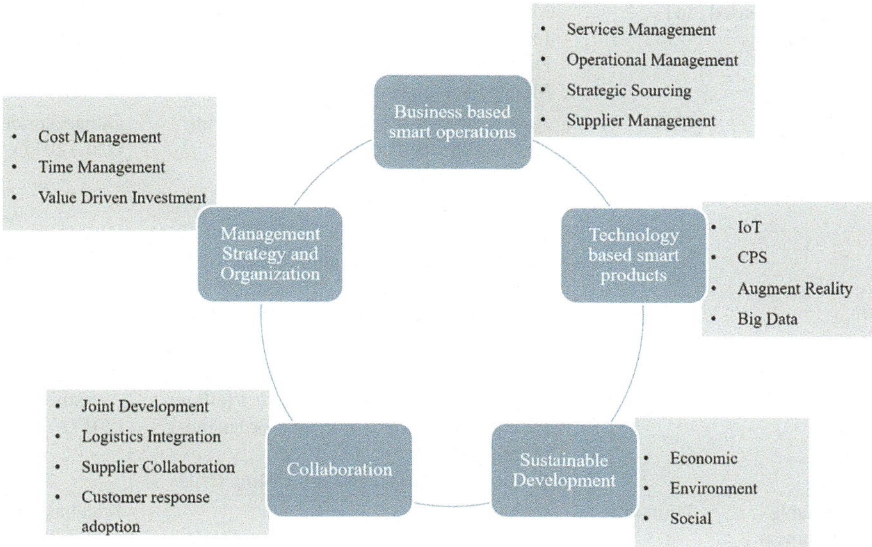

Fig. 11.1 Framework for assessing for Industry 4.0 and circularity transformation

Table 11.1 Framework for assessing Industry 4.0 and Circularity transformation

Enabler	Criteria	Attribute	Source
Business-based smart operations perspective	Services management	Streamline and monitor the execution of service Track services performance Achieve a complete view of services spend to know if it is complying	Kathawala and Abdou (2003)
	Operational management	Automate routine work/transactions Align operations to business needs Respond to exceptions before the business is affected Manage spend and reduce waste	Li and Chen (2019)
	Strategic sourcing	Maximize realized savings Get the right suppliers and right materials Structure negotiations for quality outcomes Enforce appropriate contract verbiage	Christopher and Holweg (2017)
	Supplier management	Simplify and automate supplier audits Analyze supplier performance Orchestrate supplier evaluation by various departments	Ataseven and Nair (2017)

(continued)

Table 11.1 (continued)

Enabler	Criteria	Attribute	Source
Technology-based smart products perspective	IoT	Monitor entire supply chain network Introduce cyber security Detect the environmental risks upfront Manage resource and systems	Kache and Seuring (2017)
	CPS	Assess the real-time information sharing Self-monitor and control the process Predict actions or needs of users Self-organizing production	Gunasekaran et al. (2017)
	Augment reality	Emergency management Remote assistance and guidance in maintenance activities New ways of coordination of design and manufacturing process	D'Amico et al. (2017)
	Big data	Proactive risk alerts based on historical data Deplete quality issues and product failure Flexible in consolidating data for business intelligence Predictive analytics and forecasting tools	Sayyadi and Awasthi (2017)
Sustainable development perspective	Economic	Financial stability Study risk exposure across the organization Investment in technological improvements Retrofitting approach to extend the equipment life before considering new equipment	Kocabasoglu et al. (2007)
	Environment	Reduce non-renewable energy resources Global warming with air pollution Release of water and soil pollutants	Bag et al. (2021)
	Social	Enforce health and safety practices Adhere to government regulations Awareness on macro-economics	Mani et al. (2016)
Collaboration perspective	Joint development	Collaborate across lines of business Usage of product data management (PDM) systems Reduction of non-value-added costs	Niesten et al. (2017)
	Logistics integration	Visibility on in-transit consignment location, status, and allocations Remove obstacles impacting delivery or cost variation Leverage technology & enable information-driven decision-making.	Mourtzis and Vlachou (2016)
	Supplier collaboration	Improve the cost performance index by ensuring the right material Establish collaboration with suppliers using online systems Bring the visibility to Strategic Initiatives Leverage long term supplier relationship	Canzaniello et al. (2017) and Müller et al. (2024)
	Customer response adoption	Customer satisfaction – Cost, Quality, Timeliness User experience—universal adoption, maximize savings Reduce go-to-market time by automating processes	Monostori et al. (2016)

(continued)

Table 11.1 (continued)

Enabler	Criteria	Attribute	Source
Management strategy and organization perspective	Cost management	Improve cash flow predictability and visibility throughout the supply chain Monitor price-movements of components that impact profitability Adapt to changing spending patterns Manage, process, the remarket products and services	Tarafdar and Qrunfleh (2017)
	Time management	Automate Expensive Data Management Ensure compliance for procuring timely services Mobile-based communication systems	Pagell and Wu (2017) and Sauer and Seuring (2017)
	Value-driven investment	Generate demand for goods and services Manage supply base across multiple product categories Anticipate risks and provide mitigation plans	Yawar and Seuring (2017)

The organization produces eco-friendly products, and the management is taking several measures to protect the environment in line with its commitment to sustainable development. In its choice of raw materials, sources of energy, and production processes, as well as in product development, the objective is to minimize the pollution load on the environment. In all its business operations, the organization takes serious care of social, economic, and ecological considerations. It is accredited with ISO14001:2004 certification for an effective environmental management system.

The paper organization is a right fit for this case study as the management is optimistic about developing a roadmap for applying circular supply chain practices as well as deploying new technologies by leveraging Industry 4.0. The organization is interested in implementing an ecologically sound approach toward waste minimization and cost reduction through the adoption of cleaner production technologies. The management is working with the concept of 'Waste to wealth' and encourages employees to come up with ideas to reduce the waste. The experts have hands-on experience in supply chain operations, and they are encouraged to perform the analysis. Moreover, the organization is actively involved in Corporate Social Responsibility (CSR) where every year at least 2% of the profit of the previous 3-year average will be spent on CSR activities.

11.2.1 Linear Supply Chain Before Circularity

The organization has taken a few initiatives to go green in certain aspects, indicating that they are transitioning towards a circular economy. Earlier, the organization's supply chain economy had been linear, meaning that raw materials used to make the finished goods were under-utilized. A typical linear supply chain economy of Case Organization is illustrated in Fig. 11.2.

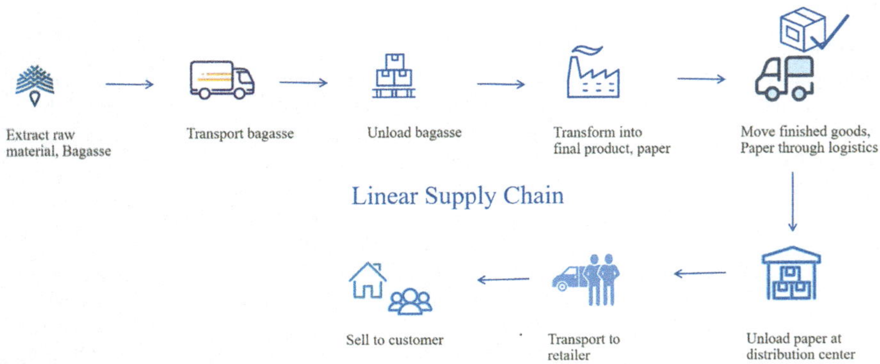

Fig. 11.2 Linear supply chain of the present organization

The supply chain of paper manufacturing is mostly straightforward and linear, but it uses renewable raw materials. For example, one of the primary raw materials used for manufacturing paper is fibrous content 'bagasse'. This raw material, bagasse is a by-product of the sugar factory. Bagasse is loaded into the truck, transported to the paper manufacturing plant, unloaded, and stored in the yard. The raw materials are processed to make pulp. The pulp is fermented, boiled, and bleached followed by a calendaring process. The outcome of the calendaring process is the finished product. Based on the raw material used the outcome can be writing paper, newsprint, or cartons. The finished goods are transported to the distribution center through third-party logistics. From the distribution center, finished goods are supplied to the retailer based on the order received. Later, the retailer sells the writing paper and newsprint to consumers.

11.2.2 *Initial Assessment of the Paper Manufacturing Organization Using MGFA*

The assessment is carried out in a paper manufacturing organization using MGFA. The framework model for evaluating the ICI is presented in Sect. 11.1.1, Table 11.1. The enablers, criteria, and attributes of the supply chain organization are represented by ICIE, ICIC, and ICIA respectively as shown in Table 11.2. Initially, a cross-functional team consisting of five Subject Matter Experts (SME) contributed to rating the factors and providing Promotional Factor (PF) as shown in Table 11.2. Further, primary assessment calculation, secondary assessment calculation, and tertiary assessment calculation are computed.

Table 11.2 Evaluation of ICI factors for paper manufacturing organization

$ICIE_i$	$ICIC_{ij}$	$ICIA_{ijk}$	SME1	SME2	SME3	SME4	SME5	PF_{ij}	PF_i	PF
$ICIE_1$	$ICIC_{11}$	$ICIA_{111}$	8	6	9	8	7	0.20	0.20	0.20
		$ICIA_{112}$	7	9	8	8	6	0.30		
		$ICIA_{113}$	7	6	8	7	8	0.50		
	$ICIC_{12}$	$ICIA_{121}$	8	6	7	8	8	0.20	0.20	
		$ICIA_{122}$	7	8	5	8	6	0.20		
		$ICIA_{123}$	5	6	5	7	5	0.40		
		$ICIA_{124}$	8	6	7	8	6	0.20		
	$ICIC_{13}$	$ICIA_{131}$	8	6	7	9	8	0.20	0.30	
		$ICIA_{132}$	6	5	5	6	5	0.40		
		$ICIA_{133}$	6	8	9	6	7	0.20		
		$ICIA_{134}$	9	6	8	7	8	0.20		
	$ICIC_{14}$	$ICIA_{141}$	8	6	7	8	9	0.30	0.30	
		$ICIA_{142}$	7	6	7	8	6	0.20		
		$ICIA_{143}$	8	6	7	7	8	0.50		
$ICIE_2$	$ICIC_{21}$	$ICIA_{211}$	7	7	6	7	8	0.40	0.30	0.30
		$ICIA_{212}$	8	6	8	5	7	0.10		
		$ICIA_{213}$	7	6	7	9	8	0.30		
		$ICIA_{214}$	6	9	5	8	7	0.20		
	$ICIC_{22}$	$ICIA_{221}$	8	7	6	8	7	0.40	0.40	
		$ICIA_{222}$	9	6	8	6	7	0.10		
		$ICIA_{223}$	8	9	6	7	5	0.30		
		$ICIA_{224}$	9	7	6	8	7	0.20		
	$ICIC_{23}$	$ICIA_{231}$	8	6	8	7	8	0.30	0.20	
		$ICIA_{232}$	6	5	7	7	8	0.30		
		$ICIA_{233}$	9	8	6	8	7	0.40		
	$ICIC_{24}$	$ICIA_{241}$	5	6	5	6	6	0.40	0.10	
		$ICIA_{242}$	9	8	7	8	6	0.10		
		$ICIA_{243}$	6	7	8	6	5	0.30		
		$ICIA_{244}$	9	6	8	9	7	0.20		
$ICIE_3$	$ICIC_{31}$	$ICIA_{311}$	8	7	5	8	8	0.40	0.40	0.20
		$ICIA_{312}$	9	8	7	8	6	0.10		
		$ICIA_{313}$	6	5	5	6	5	0.30		
		$ICIA_{314}$	9	6	8	9	7	0.20		
	$ICIC_{32}$	$ICIA_{321}$	5	6	6	5	6	0.30	0.30	
		$ICIA_{322}$	6	8	7	8	8	0.40		
		$ICIA_{323}$	6	5	5	5	6	0.30		
	$ICIC_{33}$	$ICIA_{331}$	6	6	7	7	8	0.40	0.30	
		$ICIA_{332}$	8	7	6	9	7	0.30		
		$ICIA_{333}$	9	6	7	8	7	0.30		
$ICIE_4$	$ICIC_{41}$	$ICIA_{411}$	8	6	7	8	9	0.30	0.30	0.20
		$ICIA_{412}$	7	6	7	8	6	0.20		

(continued)

Table 11.2 (continued)

$ICIE_i$	$ICIC_{ij}$	$ICIA_{ijk}$	SME1	SME2	SME3	SME4	SME5	PF_{ij}	PF_i	PF
		$ICIA_{413}$	7	6	7	7	6	0.50		
	$ICIC_{42}$	$ICIA_{421}$	6	6	6	7	8	0.40	0.30	
		$ICIA_{422}$	7	7	6	8	7	0.30		
		$ICIA_{423}$	9	6	7	8	7	0.30		
	$ICIC_{43}$	$ICIA_{431}$	7	6	7	8	6	0.40	0.20	
		$ICIA_{432}$	6	7	8	7	9	0.30		
		$ICIA_{433}$	8	6	8	9	7	0.10		
		$ICIA_{434}$	9	8	8	9	7	0.20		
	$ICIC_{44}$	$ICIA_{441}$	7	6	6	7	5	0.40	0.20	
		$ICIA_{442}$	8	7	6	9	7	0.30		
		$ICIA_{443}$	9	6	7	8	7	0.30		
$ICIE_5$	$ICIC_{51}$	$ICIA_{511}$	8	7	5	7	8	0.40	0.50	0.10
		$ICIA_{512}$	8	6	8	5	7	0.10		
		$ICIA_{513}$	8	6	7	9	9	0.30		
		$ICIA_{514}$	6	9	5	8	7	0.20		
	$ICIC_{52}$	$ICIA_{521}$	9	6	9	7	8	0.40	0.20	
		$ICIA_{522}$	7	6	6	7	6	0.40		
		$ICIA_{523}$	6	5	6	5	5	0.20		
	$ICIC_{53}$	$ICIA_{531}$	6	6	8	7	5	0.40	0.30	
		$ICIA_{532}$	7	7	6	8	7	0.30		
		$ICIA_{533}$	8	6	7	8	7	0.30		

11.2.2.1 Primary Assessment Calculation

The calculation related to the 'Services Management' criterion is represented as follows:

PF of 'Services Management' criterion, PF_{11}= (0.2,0.3,0.5)

The evaluation vector related to the 'Services Management' criterion is represented by

$$A_{11} = \begin{bmatrix} 8 & 6 & 9 & 8 & 7 \\ 7 & 9 & 8 & 8 & 6 \\ 7 & 6 & 8 & 7 & 8 \end{bmatrix}$$

The ICI of an Organization is the product of overall assessment factor A and promotional factor PF.

ICI of 'Services Management' criterion is represented by

$$ICI_{11} = PF11 \times A_{11}$$
$$ICI_{11} = (7.20,\ 6.90,\ 8.20,\ 7.50,\ 7.20)$$

Similarly, the ICI for the remaining criteria is calculated as follows:

$$
\begin{aligned}
ICI_{12} &= (6.60,\ 6.40,\ 5.80,\ 7.60,\ 6.00) \\
ICI_{13} &= (7.00,\ 6.00,\ 6.80,\ 6.80,\ 6.60) \\
ICI_{14} &= (7.80,\ 6.00,\ 7.00,\ 7.50,\ 7.90) \\
ICI_{21} &= (6.90,\ 7.00,\ 6.30,\ 7.60,\ 7.70) \\
ICI_{22} &= (8.30,\ 7.50,\ 6.20,\ 7.50,\ 6.40) \\
ICI_{23} &= (7.80,\ 6.50,\ 6.90,\ 7.40,\ 7.60) \\
ICI_{24} &= (6.50,\ 6.50,\ 6.70,\ 6.80,\ 5.90) \\
ICI_{31} &= (7.70,\ 6.30,\ 5.80,\ 7.60,\ 6.70) \\
ICI_{32} &= (5.70,\ 6.50,\ 6.10,\ 6.20,\ 6.80) \\
ICI_{33} &= (7.50,\ 6.30,\ 6.70,\ 7.90,\ 7.40) \\
ICI_{41} &= (7.30,\ 6.00,\ 7.00,\ 7.50,\ 6.90) \\
ICI_{42} &= (7.20,\ 6.30,\ 6.30,\ 7.60,\ 7.40) \\
ICI_{43} &= (7.20,\ 6.70,\ 7.60,\ 8.00,\ 7.20) \\
ICI_{44} &= (7.90,\ 6.30,\ 6.30,\ 7.90,\ 6.20) \\
ICI_{51} &= (7.60,\ 7.00,\ 5.90,\ 7.60,\ 8.00) \\
ICI_{52} &= (7.60,\ 5.80,\ 7.20,\ 6.60,\ 6.60) \\
ICI_{53} &= (6.90,\ 6.30,\ 7.10,\ 7.60,\ 6.20)
\end{aligned}
$$

11.2.2.2 Secondary Assessment Calculation

The calculation related to the 'Business Perspective' enabler is represented as follows:

PF of 'Business Perspective' enabler, $PF_1 = (0.2, 0.2, 0.3, 0.3)$

The evaluation vector related to the 'Business Perspective' enabler is represented by

$$
A_1 = \begin{bmatrix} 7.20 & 6.90 & 8.20 & 7.50 & 7.20 \\ 6.60 & 6.40 & 5.80 & 7.60 & 6.00 \\ 7.00 & 6.00 & 6.80 & 6.80 & 6.60 \\ 7.80 & 6.00 & 7.00 & 7.50 & 7.90 \end{bmatrix}
$$

ICI of 'Business Perspective' enabler is represented by

$$
\begin{aligned}
ICI_1 &= PF1 \times A1 \\
ICI_1 &= (7.20,\ 6.26,\ 6.94,\ 7.31,\ 6.99)
\end{aligned}
$$

Similarly, the ICI for remaining enablers is calculated as follows:

$$\begin{aligned}
ICI_2 &= (7.60, 7.05, 6.42, 7.44, 6.98) \\
ICI_3 &= (7.04, 6.36, 6.16, 7.27, 6.94) \\
ICI_4 &= (7.37, 6.29, 6.77, 7.71, 6.97) \\
ICI_5 &= (7.39, 6.55, 6.52, 7.40, 7.18)
\end{aligned}$$

11.2.2.3 Tertiary Assessment Calculation

The value of ICI of the paper manufacturing organization is calculated as follows:

Overall Promotional Factor, PF = (0.2, 0.3, 0.2, 0.2, 0.1)

Overall Assessment vector,

$$A = \begin{bmatrix} 7.20 & 6.26 & 6.94 & 7.31 & 6.99 \\ 7.60 & 7.05 & 6.42 & 7.44 & 6.98 \\ 7.04 & 6.36 & 6.16 & 7.27 & 6.94 \\ 7.37 & 6.29 & 6.77 & 7.71 & 6.97 \\ 7.39 & 6.55 & 6.52 & 7.40 & 7.18 \end{bmatrix}$$

$$\begin{aligned}
ICI &= PF \times A \\
ICI &= (7.34, 6.55, 6.55, 7.43, 6.99) \\
ICI &= \frac{1}{S}(7.34, 6.55, 6.55, 7.43, 6.99)
\end{aligned}$$

Where 'S' is the number of subject matter experts evaluated.

$$ICI = 6.97 \epsilon (6, 8)$$

Initial ICI for paper manufacturing organization = 6.97

11.2.2.4 Improvement Areas for the Paper Manufacturing Organization

The computation of ICI using MGFA is 6.97. This is in the range (6–7) which shows that the paper organization is slightly lagging in implementing Industry 4.0 and circularity strategies. There is scope for improvement in the paper's organization. The areas found for ICI improvements in the paper organization are described as follows:

- Respond to exceptions before the business is affected ($ICIA_{123}$)
- Get the right suppliers and the right materials ($ICIA_{132}$)
- Proactive risk alerts based on historical data ($ICIA_{241}$)
- Investment in technological improvements ($ICIA_{313}$)
- Reduce non-renewable energy resources ($ICIA_{321}$)
- Release of water and soil pollutants ($ICIA_{323}$)
- Mobile-based communication systems ($ICIA_{523}$)

11.2.3 Proposed Toolkit for the Paper Manufacturing Organization

With the identified improvement areas in the paper manufacturing organization, there are specific Industry 4.0 tools and Circularity suggested to the organization. As illustrated in Fig. 11.2, the toolkit is implemented for the identified improvement areas in the pump supply chain. The study is conducted from cradle to cradle, including their supply chain partners to transform the present linear supply chain towards CSC. The research focuses on the current industrial process and opportunities available with Industry 4.0 tools and Circularity strategies such as 6Rs to improve a synergetic network.

As illustrated in Fig. 11.3, a circular supply chain can be implemented in the paper manufacturing organization. When the pulp is fermented, a residue is formed which is used as an ingredient in the spirit manufacturing process. Currently, this residue is supplied to local spirit manufacturers. However, if the packing of the residue is done with high quality and supplied then there is an opportunity to supply the

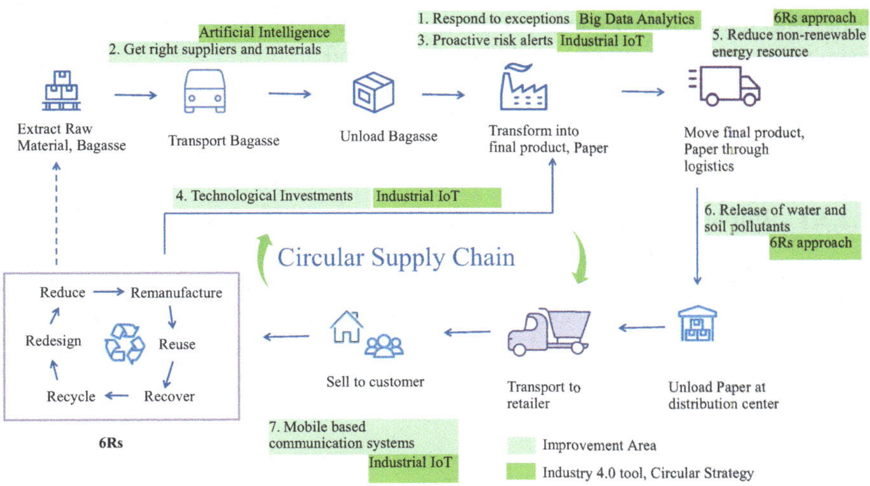

Fig. 11.3 Proposed toolkit for the paper manufacturing organization

Table 11.3 Toolkit for the Paper Manufacturing Organization

Perspective	Identified improvement areas	Industry 4.0 tool/ Circularity
Business	Respond to exceptions before the business is affected	Big data analytics
Business	Get the right suppliers and right materials	Artificial Intelligence
Technology	Proactive risk alerts based on historical data	Industrial IoT
Sustainable development	Investment in technological improvements	Industrial IoT
Sustainable development	Reduce non-renewable energy resource	6Rs approach
Sustainable development	Release of water and soil pollutants	6Rs approach
Management	Mobile-based communication systems	Industrial IoT

residue at a relatively higher cost for a potential global spirit manufacturer. After the fermentation process, the pulp is boiled at a high temperature. The wastepaper collected from the wastepaper recycles factory is used as a raw material as well during the boiling process. Coal is used for boiling the pulp. The outcome of boiling the pulp is semi-finished paper. During the boiling process, excessive steam is generated. The generated steam can be used to produce electricity, which can be utilized within the plant for power consumption as well as to cook food for employees. A large amount of coal is used to boil the pulp which results in a volume of fly ash. The fly ash can be supplied to the cement industry, as a bunch of cement manufacturers are located in the same region. Once the boiling of the pulp is completed, bleaching is carried out to decide the color of the paper. During this process, a huge amount of water is used. The used water consists of sludge and waste materials. The wastewater can be treated which can be used for irrigation purposes for nearby farms. The semi-finished paper is further processed with the calendaring process where the size of the paper is decided as per demand.

The toolkit for integration of Industry 4.0 and Circularity for the paper manufacturing organization is illustrated in Table 11.3.

11.2.3.1 Implementation of the Toolkit

The organization has an encouraging environment to implement Industry 4.0 tools and Circular practices. The stakeholders are interested in digital and circular transformation. The specific Industry 4.0 tools and Circularity implemented for the paper manufacturing organization are illustrated in Table 11.4.

Table 11.4 Implementation of the toolkit for the paper manufacturing organization

Identified improvement areas	Industry 4.0 tool/ Circularity	Implementation
Respond to exceptions before the business is affected	Big data analytics	The supply of sugarcane fluctuates based on seasonal demands and hence the supply of bagasse fluctuates. This makes paper manufacturers' plans difficult for seamless paper production. In that case, an alternate raw material like softwood is used. Big Data analytics helps the stakeholder to make the right decision with earlier supply patterns.
Get the right suppliers and right materials	Artificial Intelligence	The boiling process time depends on the grade of coal used. Based on the grade of coal, the lighting point varies. When the lighting point of coal increases, the results turn out to be improper burning with more fly ash. In other words, less lighting points results in a quick boil of pulp and less fly ash. Artificial intelligence is used to classify the coal based on moisture content in coal.
Proactive risk alerts based on historical data	Industrial IoT	The duration of fermentation varies based on the moisture content in the bagasse. Industrial IoT is used to monitor the moisture content in bagasse and alert the stakeholders when it reaches the threshold limit. This results in reduced fermentation duration and overall manufacturing lead time.
Investment in technological improvements	Industrial IoT	There is a need for monitoring key indicators remotely such as temperature variations and unusual patterns of boiling time. Industrial IoT helps to pre-detect the issue and alerts the stakeholder on the variations.
Reduce non-renewable energy resource	6Rs approach	*1. Recover generated energy within the firm* Pumps are used to move water, pulp, etc. With the help of advanced technology such as IoT, the stakeholders are informed about worn impellers, mechanical seals on pumps. Replacing the items at the right time improves the equipment efficiency and saves energy. *2. Reduce Production wastes* Felt dryer cylinders are used to press the sheets and remove the water content. It is observed that felt dryer cylinders are not insulated properly resulting in heat loss. Using IoT, the dryer cylinder can be monitored in real-time and the loss can be avoided.
Release of water and soil pollutants	6Rs approach	During the bleaching process, a huge amount of water is used. The used water consists of sludge and waste materials. The wastewater is treated that can be used for internal domestic and irrigation purposes for the nearby farms.

(continued)

Table 11.4 (continued)

Identified improvement areas	Industry 4.0 tool/ Circularity	Implementation
Mobile-based communication systems	Industrial IoT	IoT-enabled wastepaper collection in bins at customer place is introduced. It helps the organization to promptly collect the shredded papers and recycle them. IoT-enabled system will monitor the bin capacity and alerts to the organization and any third-party organization can be arranged to collect the waste papers. This improves mobile-based communications across the supply chain.

11.2.4 Final Assessment of the Paper Manufacturing Organization Using MGFA

The final assessment is carried out in the paper manufacturing organization similar to the pump manufacturing organization using MGFA. The developed framework for evaluation is assessed again in the paper manufacturing organization post-implementation of the toolkit with the subject matter experts. For the final assessment, five experts from five different departments are selected in the respective organization with more than 10 years of professional experience in the supply chain. The experts are asked to evaluate the Industry 4.0 and Circularity factors based on the Likert's Scale. Based on the score, the primary assessment is calculated as illustrated in Table 11.5. Similarly, the index for remaining Industry 4.0 and Circularity factors are calculated. Further, secondary, and tertiary assessments are carried out alike. Based on the computation, Industry 4.0 and Circularity Index (ICI) are derived.

11.2.4.1 Primary Assessment Calculation

The calculation related to the 'Services Management' criterion has shown below:

PF of 'Services Management' criterion, $PF_{11} = (0.2,0.3,0.5)$

Evaluation vector related to 'Services Management' criterion is represented by

$$A_{11} = \begin{bmatrix} 10 & 8 & 9 & 9 & 8 \\ 10 & 9 & 8 & 9 & 10 \\ 8 & 9 & 8 & 9 & 10 \end{bmatrix}$$

The ICI of an Organization is the product of overall assessment factor A and promotional factor PF.

ICI of 'Services Management' criterion is represented by

$$ICI_{11} = PF11 \times A_{11}$$
$$ICI_{11} = (9.00,\ 8.80,\ 8.20,\ 9.00,\ 9.60)$$

Table 11.5 Final evaluation of ICI factors for Paper Manufacturing Organization

$ICIE_i$	$ICIC_{ij}$	$ICIA_{ijk}$	SME1	SME2	SME3	SME4	SME5	PF_{ij}	PF_i	PF
$ICIE_1$	$ICIC_{11}$	$ICIA_{111}$	10	8	9	9	8	0.20	0.20	0.20
		$ICIA_{112}$	10	9	8	9	10	0.30		
		$ICIA_{113}$	8	9	8	9	10	0.50		
	$ICIC_{12}$	$ICIA_{121}$	9	9	8	10	8	0.20	0.20	
		$ICIA_{122}$	9	8	9	8	10	0.20		
		$ICIA_{123}$	9	9	8	9	10	0.40		
		$ICIA_{124}$	8	9	9	8	9	0.20		
	$ICIC_{13}$	$ICIA_{131}$	8	8	9	9	8	0.20	0.30	
		$ICIA_{132}$	10	9	8	8	8	0.40		
		$ICIA_{133}$	9	8	9	10	9	0.20		
		$ICIA_{134}$	9	9	8	9	8	0.20		
	$ICIC_{14}$	$ICIA_{141}$	8	9	9	8	10	0.30	0.30	
		$ICIA_{142}$	7	9	7	8	6	0.20		
		$ICIA_{143}$	8	9	10	10	8	0.50		
$ICIE_2$	$ICIC_{21}$	$ICIA_{211}$	9	7	10	7	9	0.40	0.30	0.30
		$ICIA_{212}$	8	8	8	10	7	0.10		
		$ICIA_{213}$	9	10	7	9	8	0.30		
		$ICIA_{214}$	9	9	10	8	9	0.20		
	$ICIC_{22}$	$ICIA_{221}$	10	7	8	8	9	0.40	0.40	
		$ICIA_{222}$	9	10	8	9	9	0.10		
		$ICIA_{223}$	8	9	10	7	9	0.30		
		$ICIA_{224}$	9	7	9	8	9	0.20		
	$ICIC_{23}$	$ICIA_{231}$	8	10	8	10	8	0.30	0.20	
		$ICIA_{232}$	9	10	7	7	8	0.30		
		$ICIA_{233}$	9	8	6	8	10	0.40		
	$ICIC_{24}$	$ICIA_{241}$	9	10	9	8	8	0.40	0.10	
		$ICIA_{242}$	9	8	9	8	9	0.10		
		$ICIA_{243}$	10	10	8	9	10	0.30		
		$ICIA_{244}$	9	9	8	9	7	0.20		
$ICIE_3$	$ICIC_{31}$	$ICIA_{311}$	9	9	9	8	10	0.40	0.40	0.20
		$ICIA_{312}$	9	8	7	8	9	0.10		
		$ICIA_{313}$	10	9	8	9	10	0.30		
		$ICIA_{314}$	9	9	8	9	10	0.20		
	$ICIC_{32}$	$ICIA_{321}$	9	9	8	10	9	0.30	0.30	
		$ICIA_{322}$	10	8	10	9	8	0.40		
		$ICIA_{323}$	10	9	9	8	10	0.30		
	$ICIC_{33}$	$ICIA_{331}$	9	9	10	8	8	0.40	0.30	
		$ICIA_{332}$	8	9	8	9	7	0.30		
		$ICIA_{333}$	9	10	8	8	10	0.30		
$ICIE_4$	$ICIC_{41}$	$ICIA_{411}$	8	8	9	8	9	0.30	0.30	0.20
		$ICIA_{412}$	9	10	9	8	6	0.20		

(continued)

Table 11.5 (continued)

$ICIE_i$	$ICIC_{ij}$	$ICIA_{ijk}$	SME1	SME2	SME3	SME4	SME5	PF_{ij}	PF_i	PF
		$ICIA_{413}$	8	9	9	10	6	0.50		
	$ICIC_{42}$	$ICIA_{421}$	10	8	9	9	8	0.40	0.30	
		$ICIA_{422}$	7	7	10	8	7	0.30		
		$ICIA_{423}$	9	9	9	8	10	0.30		
	$ICIC_{43}$	$ICIA_{431}$	8	9	10	8	9	0.40	0.20	
		$ICIA_{432}$	9	9	8	10	9	0.30		
		$ICIA_{433}$	8	10	8	9	7	0.10		
		$ICIA_{434}$	9	8	10	9	9	0.20		
	$ICIC_{44}$	$ICIA_{441}$	10	9	9	9	9	0.40	0.20	
		$ICIA_{442}$	9	10	9	9	7	0.30		
		$ICIA_{443}$	9	9	8	10	7	0.30		
$ICIE_5$	$ICIC_{51}$	$ICIA_{511}$	8	10	9	8	8	0.40	0.50	0.10
		$ICIA_{512}$	8	9	8	5	7	0.10		
		$ICIA_{513}$	8	9	9	10	9	0.30		
		$ICIA_{514}$	9	9	10	8	10	0.20		
	$ICIC_{52}$	$ICIA_{521}$	9	9	9	10	8	0.40	0.20	
		$ICIA_{522}$	7	10	9	7	10	0.40		
		$ICIA_{523}$	9	10	9	9	8	0.20		
	$ICIC_{53}$	$ICIA_{531}$	9	8	9	9	10	0.40	0.30	
		$ICIA_{532}$	10	8	9	8	9	0.30		
		$ICIA_{533}$	8	10	9	8	10	0.30		

Similarly, the ICI for the remaining criteria is calculated as follows:

$$ICI_{12} = (8.80, 8.80, 8.40, 8.80, 9.40)$$
$$ICI_{13} = (9.20, 8.60, 8.40, 8.80, 8.20)$$
$$ICI_{14} = (7.80, 9.00, 9.10, 9.00, 8.20)$$
$$ICI_{21} = (8.90, 8.40, 8.90, 8.10, 8.50)$$
$$ICI_{22} = (9.10, 7.90, 8.80, 7.80, 9.00)$$
$$ICI_{23} = (8.70, 9.20, 6.90, 8.30, 8.80)$$
$$ICI_{24} = (9.30, 9.60, 8.50, 8.50, 8.50)$$
$$ICI_{31} = (9.30, 8.90, 8.30, 8.50, 9.90)$$
$$ICI_{32} = (9.70, 8.60, 9.10, 9.00, 8.90)$$
$$ICI_{33} = (8.70, 9.30, 8.80, 8.30, 8.30)$$
$$ICI_{41} = (8.20, 8.90, 9.00, 9.00, 6.90)$$
$$ICI_{42} = (8.80, 8.00, 9.30, 8.40, 8.30)$$
$$ICI_{43} = (8.50, 8.90, 9.20, 8.90, 8.80)$$
$$ICI_{44} = (9.40, 9.30, 8.70, 9.30, 7.80)$$
$$ICI_{51} = (8.20, 9.40, 9.10, 8.30, 8.60)$$
$$ICI_{52} = (8.20, 9.60, 9.00, 8.60, 8.80)$$
$$ICI_{53} = (9.00, 8.60, 9.00, 8.40, 9.70)$$

11.2.4.2 Secondary Assessment Calculation

The calculation related to the 'Business Perspective' enabler is represented as follows:

PF of 'Business Perspective' enabler, $PF_1 = (0.2, 0.2, 0.3, 0.3)$

Evaluation vector related to 'Business Perspective' enabler is represented by

$$A_1 = \begin{bmatrix} 9.00 & 8.80 & 8.20 & 9.00 & 9.60 \\ 8.80 & 8.80 & 8.40 & 8.80 & 9.40 \\ 9.20 & 8.60 & 8.40 & 8.80 & 8.20 \\ 7.80 & 9.00 & 9.10 & 9.00 & 8.20 \end{bmatrix}$$

ICI of 'Business Perspective' enabler is represented by

$$ICI_1 = PF1 \times A1$$
$$ICI_1 = (8.66, 8.80, 8.57, 8.90, 8.72)$$

Similarly, the ICI for remaining enablers is calculated as follows:

$$ICI_2 = (8.98, 8.48, 8.42, 8.06, 8.76)$$
$$ICI_3 = (9.24, 8.93, 8.69, 8.59, 9.12)$$
$$ICI_4 = (8.68, 8.71, 9.07, 8.86, 7.88)$$
$$ICI_5 = (8.44, 9.20, 9.05, 8.39, 8.97)$$

11.2.4.3 Tertiary Assessment Calculation

The value of ICI of the paper manufacturing organization is calculated as follows:

Overall Promotional Factor, PF= (0.2, 0.3, 0.2, 0.2, 0.1)

Overall Assessment vector,

$$A = \begin{bmatrix} 8.66 & 8.80 & 8.57 & 8.90 & 8.72 \\ 8.98 & 8.48 & 8.42 & 8.06 & 8.76 \\ 9.24 & 8.93 & 8.69 & 8.59 & 9.12 \\ 8.68 & 8.71 & 9.07 & 8.86 & 7.88 \\ 8.44 & 9.20 & 9.05 & 8.39 & 8.97 \end{bmatrix}$$

$$ICI = PF \times A$$
$$ICI = (8.85, 8.75, 8.70, 8.53, 8.67)$$
$$ICI = \frac{1}{S}(8.85, 8.75, 8.70, 8.53, 8.67)$$

Where 'S' is the number of subject matter experts evaluated.

$$\text{ICI} = 8.70\epsilon(8,9)$$

Final ICI for paper manufacturing organization = 8.70

Based on the assessment of Industry 4.0 and Circularity factors, the ICI derived using MGFA for the pump manufacturing organization is 8.7. This is in the range of (8–10) which shows that the paper manufacturing organization is exceeding on digital and circularity transformation. The results coincide with the practical environment as the paper manufacturing organization has implemented the recommended strategy.

11.2.4.4 Statistical Validation

The formulated hypothesis is represented as follows.

H0—The developed Industry 4.0 and Circularity (ICI) framework effectively assesses the digital and circularity level of the organization.
H1—The developed Industry 4.0 and Circularity (ICI) framework does not effectively assess the digital and circularity level of the organization.

The responses of the experts are presented in Table 11.6.

The statistical study is carried out to analyze the feedback received from SMEs on the feasibility of assessing the digital and circularity level using a developed ICI framework. The ratings given by SMEs are given in Statistical Package for the Social Sciences (SPSS) and a t-test is conducted to examine the acceptance of 'assessment of digital and circularity level using MGFA'. In the first instance, the test value is given as 10 which means that '100% of the opinions are in favor of successful assessment of digital and circularity level using MGFA in practice at 95% confidence interval'. The null hypothesis is not satisfied in this case.

Table 11.6 Consolidated responses of the experts

	Expert response					Average rating in Likert's scale of range 0–10
Question	SME1	SME2	SME3	SME4	SME5	
To what extent do you believe that the developed framework broadly represents all perspectives of Industry 4.0 and circularity in your organization?	9	8	9	9	9	8.8
To what extent do you believe that the multi-grade fuzzy approach is effective?	9	8	8	9	7	8.2
To what extent do you believe that the developed Industry 4.0 and circularity framework is easy to use?	8	9	8	8	9	8.4

In the second case, the null hypothesis is set as '90% of the opinions are in favor of successful assessment of digital and circularity level using MGFA in practice at 95% confidence interval'. The output of the t-test is shown in Table 11.7. In this case, the null hypothesis is accepted. On the whole, this validation study indicates the feasibility of assessing the digital and circularity level using developed ICI framework with the success rate of 90%.

11.3 Results and Discussions

The ICI for case organizations before and after implementation of Industry 4.0 tools and Circularity practices suggests that there are significant improvements in the areas identified as illustrated in Fig. 11.4. The statistical validation carried out before and after implementation revealed that the results are practical and sustainable. Increased visibility and agility provided by Industry 4.0 tools and Circularity practices have a considerable role in providing flexible, agile, robust, and sustainable supply chains.

Table 11.7 Statistical study with 95% confidence

t	df	Mean difference	95% Confidence interval of the difference	
			Lower	Upper
−1.000	2	−.33333	−1.7676	1.1009
−2.000	2	−.66667	−2.1009	.7676
−2.000	2	−.66667	−2.1009	.7676
−1.000	2	−.33333	−1.7676	1.1009
−1.000	2	−.66667	−3.5351	2.2018

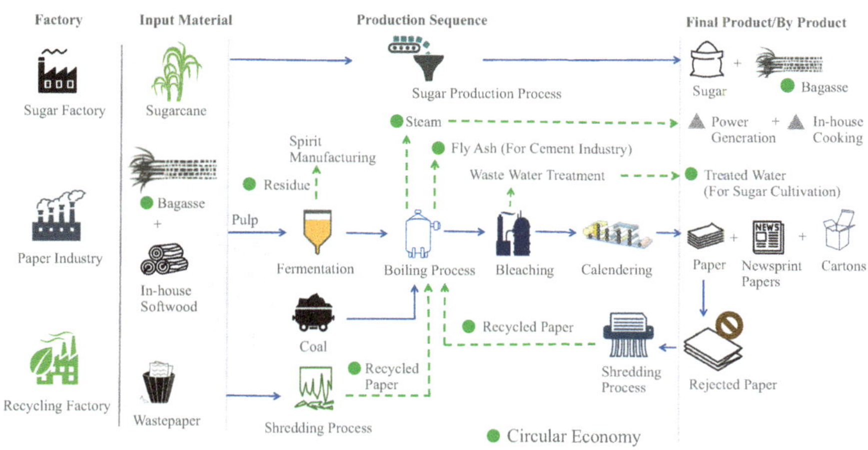

Fig. 11.4 Implementation of a toolkit for the paper manufacturing organization

11.4 Key Implementation Steps to Deploy Industry 4.0 and Circularity

Based on the study conducted, the holistic implementation steps to deploy Industry 4.0 and Circularity are illustrated in Fig. 11.5. Step 1 and Step 2 are presented from an organizational strategy perspective and Steps 3–6 are presented from a practical implementation perspective.

Step 1—Identify Improvement Areas
The roadmap to implementation commences by identifying the supply chain areas that are obstructing the overall supply chain operational efficiency. The improvement areas are identified by Key Performance Indicators (KPIs) throughout the supply chain.

Step 2—Build a Strategy
Step 2 is a crucial stage for building a suitable strategy. During this step, the organization should recommend a potential long-term solution for the identified improvement areas using the selected Industry 4.0 tools and Circularity. When selecting the right toolkit, the organization must also consider the maximum available budget with a threshold limit for the implementation (Fig. 11.6).

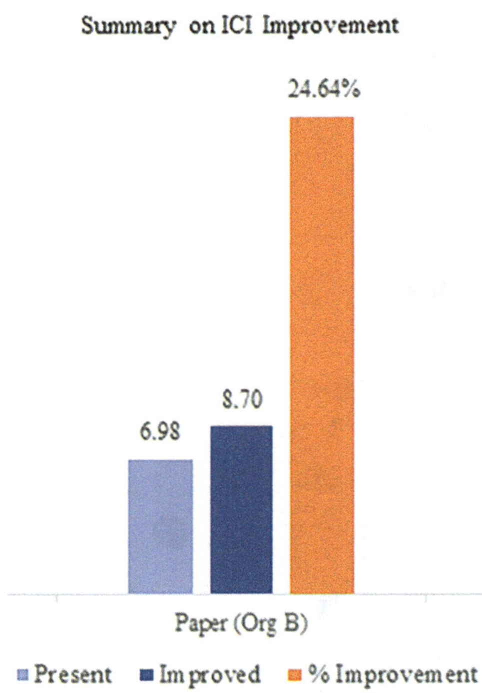

Fig. 11.5 Before and after implementing Industry 4.0 and circularity

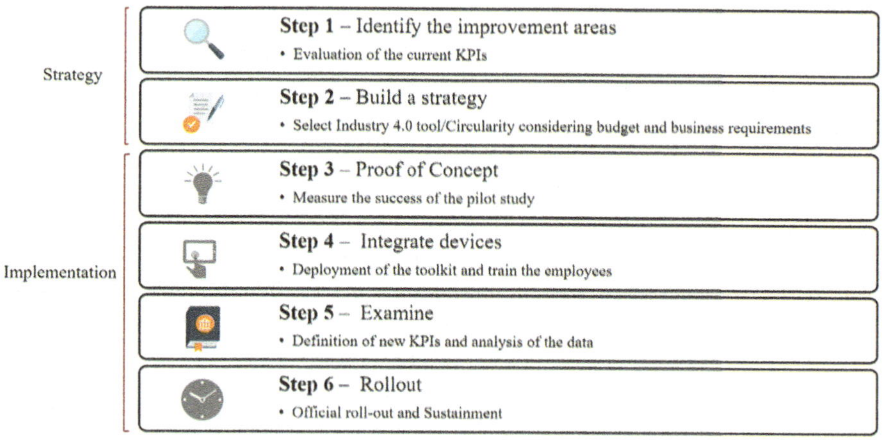

Fig. 11.6 Key implementation steps to deploy Industry 4.0 and Circularity

Step 3—Proof of Concept

The strategy begins with the initial implementation of the Industry 4.0 tools and Circularity selected as a part of the pilot study. This initial implementation begins with a Proof of Concept (POC), not by installing the tool across the entire organization. The success of the POC must be assessed during a well-defined timeline. If the POC does not prove the expected outcome, the process must go back to Step 2 to recommend a different long-term solution.

Step 4—Integrate Devices

If the outcomes during Step 3 are satisfying, the approach should be continued to deploy the solution throughout the entire business functions for the trial run. This phase begins by training the workforce in the organization on how the tool is used and by conducting dry runs. This phase will improve the acceptance of the selected Industry 4.0 tool and Circularity.

Step 5—Examine

The initial KPIs identified during Step 1 are revisited and it may not be adequate at times. During this phase, additional KPIs can be proposed based on the data gathered by integrating Information and Communication Technology (ICT) systems.

Step 6—Rollout

The final phase is the formal roll-out of Industry 4.0 and Circularity throughout the entire business functions. This phase concentrates on examining the well-defined KPIs. The project management group should involve the implementation team until hyper care support and gradually a sustainment team needs to take over the proceedings.

Thus, the proposed implementation steps to deploy Industry 4.0 and Circularity that aims to facilitate quick decision-making and improve the efficiency of overall supply chain operations.

11.5 Conclusion

In this extensive research endeavor, a thorough exploration of supply chain dynamics, Industry 4.0 tools, and Circularity practices has been undertaken through a comprehensive literature review. By synthesizing insights from industry experts and stakeholders, key factors influencing an organization's preparedness for Industry 4.0 and Circularity transformation have been identified. Building upon this foundation, a robust framework for assessing organizational readiness in these transformative realms has been meticulously developed. The selection of a case organization was guided by the maturity of their supply chain processes and their expressed willingness to embrace digital technologies and circularity strategies. Leveraging the developed framework, the research employs the Multi-Grade Fuzzy Approach (MGFA) to assess the current state of readiness for Industry 4.0 and Circularity within each chosen organization. Through this assessment, specific improvement areas unique to each case organization are discerned, paving the way for targeted enhancements.

Ensuring the practical applicability of Industry 4.0 tools and circularity practices, a rigorous statistical validation is conducted. The culmination of these efforts manifests in the proposition and subsequent implementation of a toolkit for the case organization, aligning with the principles of Industry 4.0 and Circularity. Rigorous analysis and fine-tuning of the toolkit are conducted in collaboration with the subject matter experts. The conclusive phase of the research involves a final assessment of the factors influencing Industry 4.0 and Circularity transformation within Paper industry. Validation through interaction with subject matter experts solidifies the findings. Building upon the insights gained from the assessment in these diverse sectors, a comprehensive roadmap for Industry 4.0 and Circularity transformation is proposed. This roadmap serves not only as a testament to the research outcomes but also as a practical guide for organizations across industries seeking to navigate the transformative journey towards Industry 4.0 and Circularity.

References

Ataseven C, Nair A (2017) Assessment of supply chain integration and performance relationships: a meta-analytic investigation of the literature. Int J Prod Econ 185:252–265. https://doi.org/10.1016/j.ijpe.2017.01.007

Bag S, Gupta S, Kumar S (2021) Industry 4.0 adoption and 10R advance manufacturing capabilities for sustainable development. Int J Prod Econ 231:107844. https://doi.org/10.1016/j.ijpe.2020.107844

Canzaniello A, Hartmann E, Fifka MS (2017) Intra-industry strategic alliances for managing sustainability-related supplier risks: motivation and outcome. Int J Phys Distrib Logist Manag 47(5):387–409. https://doi.org/10.1108/IJPDLM-01-2016-0034

Christopher M, Holweg M (2017) Supply chain 2.0 revisited: a framework for managing volatility-induced risk in the supply chain. Int J Phys Distrib Logist Manag 47(1):2–17. https://doi.org/10.1108/IJPDLM-09-2016-0245

D'Amico G, Nardecchia M, Panci P, Sannino F, Strumia A, Torre R, Urbano A (2017) Flavour anomalies after the RK∗ measurement. J High Energy Phys 2017(9):1–32. https://doi.org/10.1007/JHEP09(2017)010

Dey PK, Malesios C, Chowdhury S, Saha K, Budhwar P, De D (2022) Adoption of circular economy practices in small and medium-sized enterprises: evidence from Europe. Int J Prod Econ 248:108496. https://doi.org/10.1016/j.ijpe.2022.108496

Gunasekaran A, Papadopoulos T, Dubey R, Wamba SF, Childe SJ, Hazen B, Akter S (2017) Big data and predictive analytics for supply chain and organizational performance. J Bus Res 70:308–317. https://doi.org/10.1016/j.jbusres.2016.08.004

Kache F, Seuring S (2017) Challenges and opportunities of digital information at the intersection of Big Data Analytics and supply chain management. Int J Oper Prod Manag 37(1):10–36. https://doi.org/10.1108/IJOPM-02-2015-0078

Kathawala Y, Abdou K (2003) Supply chain evaluation in the service industry: a framework development compared to manufacturing. Manag Audit J 18(2):140–149. https://doi.org/10.1108/02686900310455137

Kocabasoglu C, Prahinski C, Klassen RD (2007) Linking forward and reverse supply chain investments: the role of business uncertainty. J Oper Manag 25(6):1141–1160. https://doi.org/10.1016/j.jom.2007.01.015

Li S, Chen X (2019) The role of supplier collaboration and risk management capabilities in managing product complexity. Oper Manag Res 12(3):146–158. https://doi.org/10.1007/s12063-019-00144-w

Mani V, Agarwal R, Gunasekaran A, Papadopoulos T, Dubey R, Childe SJ (2016) Social sustainability in the supply chain: construct development and measurement validation. Ecol Indic 71:270–279. https://doi.org/10.1016/j.ecolind.2016.07.007

Monostori L, Kádár B, Bauernhansl T, Kondoh S, Kumara S, Reinhart G et al (2016) Cyber-physical systems in manufacturing. CIRP Ann 65(2):621–641. https://doi.org/10.1016/j.cirp.2016.06.005

Mourtzis D, Vlachou E (2016) Cloud-based cyber-physical systems and quality of services. TQM J 28(5):704–733. https://doi.org/10.1108/TQM-10-2015-0133

Müller JM, Islam N, Kazantsev N, Romanello R, Olivera G, Das D, Hamzeh R (2024) Barriers and enablers for industry 4.0 in SMEs: a combined integration framework. IEEE Trans Eng Manag. https://doi.org/10.1109/TEM.2024.3365771

Niesten E, Jolink A, de Sousa Jabbour ABL, Chappin M, Lozano R (2017) Sustainable collaboration: the impact of governance and institutions on sustainable performance. J Clean Prod 155:1–6. https://doi.org/10.1016/j.jclepro.2016.12.085

Pagell M, Wu Z (2017) Business implications of sustainability practices in supply chains. In: Sustainable supply chains: a research-based textbook on operations and strategy. Springer, Cham, pp 339–353. https://doi.org/10.1007/978-3-319-29791-0_15

Sauer PC, Seuring S (2017) Sustainable supply chain management for minerals. J Clean Prod 151:235–249. https://doi.org/10.1016/j.jclepro.2017.03.049

Sayyadi R, Awasthi A (2017) A system dynamics based simulation model to evaluate regulatory policies for sustainable transportation planning. Int J Model Simul 37(1):25–35. https://doi.org/10.1080/02286203.2016.1219806

Tarafdar M, Qrunfleh S (2017) Agile supply chain strategy and supply chain performance: complementary roles of supply chain practices and information systems capability for agility. Int J Prod Res 55(4):925–938. https://doi.org/10.1080/00207543.2016.1203079

Yawar SA, Seuring S (2017) Management of social issues in supply chains: a literature review exploring social issues, actions and performance outcomes. J Bus Ethics 141(3):621–643. https://doi.org/10.1007/s10551-015-2719-9

Chapter 12
Exploring Circular Supply Chain Practices and Industry 4.0 Integration: A Case Study in a Pump Manufacturing Organization

Abstract Circular supply chains are gaining more and more attention because they help reduce resource waste and increase resource utilisation efficiency. Data is the driving force behind the current industrial revolution, or Industry 4.0. It is highly anticipated that Industry 4.0 tools and technologies would improve the performance of businesses in the circular economy (CE). Many Industry 4.0 tools like big data, IIoT and Artificial intelligence are integrated in most of the supply chains to improve the circularity. Both academics and business people are paying close attention to Industry 4.0. Both supply chain performance and industry efficiency have been greatly improved by supply chain integration. A pump manufacturing organization is taken as a case study to understand the supply chain practices that have been followed by the industry by integrating Industry 4.0 tools. This chapter aims to provide a thorough and methodical understanding of circular supply chain processes with a pump manufacturing industry case study.

Keywords Circular supply chain · Industry 4.0 · Circular economy · Industrial revolution · Case study · IIoT

12.1 Introduction

The pump manufacturing industry historically entangled in a linear "take-make-dispose" model, now faces a crossroads. Rising environmental concerns and evolving customer demands call for a radical shift towards a more sustainable future (Afeltra et al. 2023). This book chapter explains this transformation, using Industry 4.0 and Circular Supply Chain Management (CSCM) within the context of a leading South Indian pump manufacturer—Case Organization. To translate theory into practice, a case organization was chosen to embrace digital transformation. Using the Multi-Grade Fuzzy Approach (MGFA) as a tool, the chapter facilitates assessment of the organization's current state of readiness. Later the chapter reveals the toolkits tailored for each case organization. These toolkits seamlessly integrate

J. Kandasamy et al., *Circular Supply Chain Management*, Circular Economy and Sustainability, https://doi.org/10.1007/978-3-031-81750-2_12

Industry 4.0 tools and Circular Supply Chain Management practices, precisely addressing the distinctive needs and challenges within each sector.

This chapter unravels the complexities of organizational preparedness for the transformative amalgamation of Industry 4.0 and Circularity and also chapter provides a comprehensive roadmap for Industry 4.0 and Circularity transformation. Drawing upon lessons distilled from diverse industries, the roadmap unfolds as a valuable guide, furnishing actionable insights for organizations navigating the transformative journey toward a technologically advanced and circular future. As we progress through the subsequent chapters, the narrative deepens, contributing to a collective understanding of the unfolding industrial evolution. This chapter offers a clear path for any organization seeking to navigate the transformative journey toward a sustainable and technologically advanced future.

12.2 About Case Organization—Pump Manufacturing Industry

Case organization is a leading pump manufacturing organization located near the biggest economic corridor in the South of India, the city of Chennai. This pump manufacturing facility utilizes modern technologies to provide clients with a perfect fit for their pumping operations in any market application and hydraulic configuration. The facility is fully equipped to manufacture, assemble, and test standard centrifugal and reciprocating pump equipment and highly engineered pump systems. The organization creates pumping solutions for the Oil and Gas, Chemical, Power, and Industrial sectors. The case organization is also developing cost-effective solutions with minimized energy consumption, high efficiency, low maintenance, and a long product lifecycle. It guarantees the exceptional quality of pumps and fast delivery of spare parts. Additionally, this plant enables the organization to maintain a reliable supply of spare parts for any pump, regardless of its manufacturer, and efficient support for customers in the Middle East, Indian, and Southeast Asian markets.

The organization has put into practice world-class approaches such as ISO 9001:2015 for quality management systems in the supply chain. The stakeholders are interested in digital transformation when approached for a case study. Moreover, as the organization is matured in the supply chain, the experts are well experienced in SCM with expertise in production and operations management, risk management, and environmental safety. The experts are interested in providing more inputs and they are easily accessible for multiple rounds of discussion. The organization thus selected is best suited for the study as they are positive about adopting new technologies and distinguish themselves as a global leader in pump and water technology.

12.2.1 Pump Supply Chain

The case organization is developing cost-effective solutions with minimized energy consumption, high efficiency, low maintenance, and a long product lifecycle. It guarantees the exceptional quality of pumps and fast delivery of spare parts. A typical pump supply chain of Case Organization A is illustrated in Fig. 12.1.

The supply chain of pump manufacturing is mostly linear. The organization procures casting components, mechanical seals, and electric motors from multiple suppliers. In addition, the purchasing department is involved in equipment purchases. The manufacturing unit consists of a machining shop, grinding shop, assembly shop, and painting shop. It enables the manufacturing process of castings, shafts, and pumps. The pumps manufactured are both engineered and standard pumps. Later finished goods are moved to the packaging unit where the assembly and packaging of products are monitored. There are few control systems facilitated to maintain quality and minimize delays in operations. Finally, the products are moved to a warehouse and third-party logistics is used to move the products to multiple distribution centers to serve the customers.

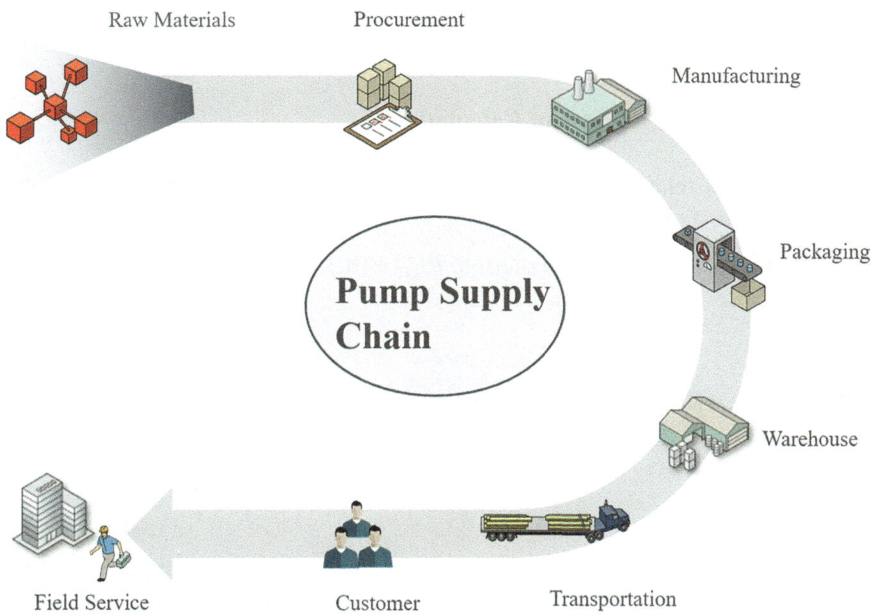

Fig. 12.1 Pump supply chain

12.2.2 Initial Assessment of Pump Manufacturing Organization Using MGFA

The assessment is primarily based on five enablers. As illustrated in Fig. 12.2, the enablers are classified into five major categories such as business, technology, sustainable development, collaboration, and management strategy. These enablers aid organizations in assessing their readiness for Industry 4.0 and Circularity transformation using MGFA. The MGFA is a widely used technique in the manufacturing and logistics sectors for the assessment of sustainability, leanness and agility (Vimal et al. 2021). MGFA is used so that it is easy for organizations to adapt and identify the improvement areas quickly. It is also used to avoid the fluctuation of respondent value and easy computations (Fosci and Psaila 2024). In MGFA, the responses are assessed in grades (extremely strong, strong, generally strong, weak, and extremely weak) and the grade is reduced gradually. Hence the assessment is fuzzy. For convenient calculation, the assessment is mentioned quantitively.

The framework model for evaluating the ICI is presented in Chap. 11, Table 11.1. The enablers, criteria, and attributes of the supply chain organization are represented by ICIE, ICIC, and ICIA respectively as shown in Table 12.1. A cross-functional team consisting of five Subject Matter Experts (SME) has contributed to rating the factors and providing Promotional Factor (PF).

12.2.2.1 Primary Assessment Calculation

The calculation related to the 'Services Management' criterion is represented as follows:

PF of 'Services Management' criterion, $PF_{11} = (0.3, 0.3, 0.4.)$

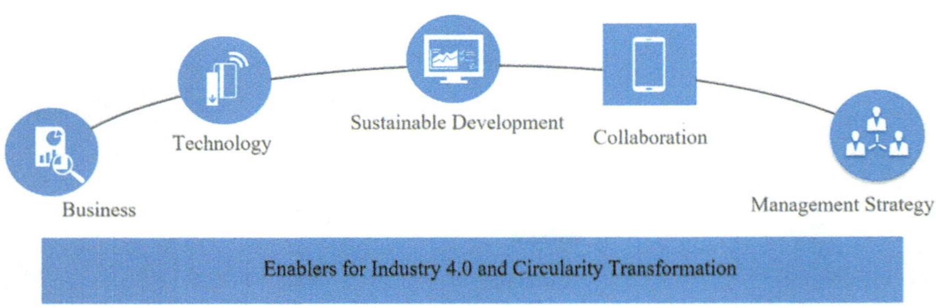

Fig. 12.2 Enablers of a supply chain organization

Table 12.1 Evaluation of ICI factors for pump manufacturing organization

$ICIE_i$	$ICIC_{ij}$	$ICIA_{ijk}$	SME1	SME2	SME3	SME4	SME5	PF_{ij}	PF_i	PF
$ICIE_1$	$ICIC_{11}$	$ICIA_{111}$	8	6	8	7	9	0.3	0.20	0.30
		$ICIA_{112}$	6	5	8	7	8	0.3		
		$ICIA_{113}$	9	8	6	8	7	0.4		
	$ICIC_{12}$	$ICIA_{121}$	8	9	6	7	6	0.2	0.40	
		$ICIA_{122}$	9	8	6	7	8	0.1		
		$ICIA_{123}$	5	6	7	6	5	0.4		
		$ICIA_{124}$	9	8	7	8	7	0.3		
	$ICIC_{13}$	$ICIA_{131}$	9	8	6	7	8	0.3	0.20	
		$ICIA_{132}$	8	7	6	6	9	0.4		
		$ICIA_{133}$	7	8	6	9	7	0.2		
		$ICIA_{134}$	9	8	7	5	6	0.1		
	$ICIC_{14}$	$ICIA_{141}$	5	6	5	7	5	0.4	0.20	
		$ICIA_{142}$	8	7	6	9	7	0.3		
		$ICIA_{143}$	9	6	7	8	7	0.3		
$ICIE_2$	$ICIC_{21}$	$ICIA_{211}$	8	7	6	9	7	0.4	0.50	0.30
		$ICIA_{212}$	9	6	8	6	7	0.1		
		$ICIA_{213}$	8	9	6	7	5	0.3		
		$ICIA_{214}$	9	7	6	8	7	0.2		
	$ICIC_{22}$	$ICIA_{221}$	9	7	5	7	8	0.4	0.20	
		$ICIA_{222}$	8	6	8	5	7	0.1		
		$ICIA_{223}$	8	6	7	9	9	0.3		
		$ICIA_{224}$	6	9	5	8	7	0.2		
	$ICIC_{23}$	$ICIA_{231}$	7	9	6	7	8	0.4	0.10	
		$ICIA_{232}$	7	5	7	8	9	0.5		
		$ICIA_{233}$	6	7	8	7	9	0.1		
	$ICIC_{24}$	$ICIA_{241}$	5	6	7	5	6	0.4	0.20	
		$ICIA_{242}$	6	9	8	7	9	0.3		
		$ICIA_{243}$	9	6	8	9	7	0.1		
		$ICIA_{244}$	9	8	8	9	7	0.2		
$ICIE_3$	$ICIC_{31}$	$ICIA_{311}$	8	7	6	9	8	0.4	0.50	0.20
		$ICIA_{312}$	9	8	7	8	6	0.1		
		$ICIA_{313}$	6	5	8	6	5	0.3		
		$ICIA_{314}$	9	6	8	9	7	0.2		
	$ICIC_{32}$	$ICIA_{321}$	8	6	7	8	7	0.3	0.20	
		$ICIA_{322}$	6	8	9	8	8	0.4		
		$ICIA_{323}$	7	9	8	9	6	0.3		
	$ICIC_{33}$	$ICIA_{331}$	5	6	5	6	5	0.50	0.30	
		$ICIA_{332}$	6	9	7	8	8	0.30		
		$ICIA_{333}$	8	7	6	9	5	0.20		
$ICIE_4$	$ICIC_{41}$	$ICIA_{411}$	8	6	7	9	9	0.30	0.30	0.10
		$ICIA_{412}$	7	6	7	8	6	0.20		

(continued)

Table 12.1 (continued)

$ICIE_i$	$ICIC_{ij}$	$ICIA_{ijk}$	SME1	SME2	SME3	SME4	SME5	PF_{ij}	PF_i	PF
		$ICIA_{413}$	5	6	7	7	5	0.50		
	$ICIC_{42}$	$ICIA_{421}$	9	6	9	6	8	0.30	0.30	
		$ICIA_{422}$	6	8	9	7	6	0.20		
		$ICIA_{423}$	9	8	9	7	6	0.50		
	$ICIC_{43}$	$ICIA_{431}$	8	6	7	9	8	0.40	0.20	
		$ICIA_{432}$	7	8	5	8	6	0.20		
		$ICIA_{433}$	9	6	8	7	7	0.20		
		$ICIA_{434}$	8	6	7	9	6	0.20		
	$ICIC_{44}$	$ICIA_{441}$	6	7	9	7	8	0.30	0.20	
		$ICIA_{442}$	6	7	8	7	6	0.50		
		$ICIA_{443}$	9	8	9	7	6	0.20		
$ICIE_5$	$ICIC_{51}$	$ICIA_{511}$	8	6	7	9	8	0.40	0.40	0.10
		$ICIA_{512}$	7	9	8	8	6	0.20		
		$ICIA_{513}$	6	8	9	6	7	0.20		
		$ICIA_{514}$	9	6	8	7	8	0.20		
	$ICIC_{52}$	$ICIA_{521}$	9	6	9	7	8	0.40	0.30	
		$ICIA_{522}$	5	5	6	7	6	0.40		
		$ICIA_{523}$	9	8	9	7	6	0.20		
	$ICIC_{53}$	$ICIA_{531}$	8	6	9	9	7	0.20	0.30	
		$ICIA_{532}$	7	9	8	8	6	0.30		
		$ICIA_{533}$	9	6	8	7	9	0.50		

Evaluation vector related to 'Services Management' criterion is represented by

$$A_{11} = \begin{bmatrix} 8 & 6 & 8 & 7 & 9 \\ 6 & 5 & 8 & 7 & 8 \\ 9 & 8 & 6 & 8 & 7 \end{bmatrix}$$

The ICI of an Organization is the product of overall assessment factor A and promotional factor PF.

ICI of the 'Services Management' criterion is represented by Eq. 12.1

$$\begin{aligned} \mathrm{ICI}_{11} &= PF_{11} \times A_{11} \\ \mathrm{ICI}_{11} &= (7.8,\ 6.5,\ 7.2,\ 7.4,\ 7.9) \end{aligned} \tag{12.1}$$

Similarly, the ICI for the remaining criteria is calculated as follows:

$$
\begin{aligned}
\mathrm{ICI}_{12} &= (7.2,\ 7.4,\ 6.7,\ 6.9,\ 6.1)\\
\mathrm{ICI}_{13} &= (8.2,\ 7.6,\ 6.1,\ 6.8,\ 8)\\
\mathrm{ICI}_{14} &= (7.1,\ 6.3,\ 5.9,\ 7.9,\ 6.2)\\
\mathrm{ICI}_{21} &= (8.3,\ 7.5,\ 6.2,\ 7.9,\ 6.4)\\
\mathrm{ICI}_{22} &= (8,\ 7,\ 5.9,\ 7.6,\ 8)\\
\mathrm{ICI}_{23} &= (6.9,\ 6.8,\ 6.7,\ 7.5,\ 8.6)\\
\mathrm{ICI}_{24} &= (6.5,\ 7.3,\ 7.6,\ 6.8,\ 7.2)\\
\mathrm{ICI}_{31} &= (7.7,\ 6.3,\ 7.1,\ 8,\ 6.7)\\
\mathrm{ICI}_{32} &= (6.9,\ 7.7,\ 8.1,\ 8.3,\ 7.1)\\
\mathrm{ICI}_{33} &= (5.9,\ 7.1,\ 5.8,\ 7.2,\ 5.9)\\
\mathrm{ICI}_{41} &= (6.3,\ 6,\ 7,\ 7.8,\ 6.4)\\
\mathrm{ICI}_{42} &= (8.4,\ 7.4,\ 9,\ 6.7,\ 6.6)\\
\mathrm{ICI}_{43} &= (8,\ 6.4,\ 6.8,\ 8.4{,}7)\\
\mathrm{ICI}_{44} &= (6.6,\ 7.2,\ 8.5,\ 7,\ 6.6)\\
\mathrm{ICI}_{51} &= (7.6,\ 7,\ 7.8,\ 7.8,\ 7.4)\\
\mathrm{ICI}_{52} &= (7.4,\ 6,\ 7.8,\ 7,\ 6.8)\\
\mathrm{ICI}_{53} &= (8.2,\ 6.9,\ 2,\ 7.7,\ 7.7)
\end{aligned}
$$

12.2.2.2 Secondary Assessment Calculation

The calculation related to the 'Business Perspective' enabler is represented as follows:

PF of 'Business Perspective' enabler, $PF_1 = (0.2, 0.4, 0.2, 0.2)$

The evaluation vector related to the 'Business Perspective' enabler is represented by

$$
A_1 = \begin{bmatrix} 7.8 & 6.5 & 7.2 & 7.4 & 7.9 \\ 7.2 & 7.4 & 6.7 & 6.9 & 6.1 \\ 8.2 & 7.6 & 6.1 & 6.8 & 8.0 \\ 7.1 & 6.3 & 5.9 & 7.9 & 6.2 \end{bmatrix}
$$

ICI of the 'Business Perspective' enabler is represented by Eq. 12.2

$$
\begin{aligned}
\mathrm{ICI}_1 &= PF_1 \times A_1\\
\mathrm{ICI}_1 &= (7.5,\ 7.04,\ 6.52,\ 7.18,\ 6.86)
\end{aligned}
\tag{12.2}
$$

Similarly, the ICI for remaining enablers is calculated as follows:

$$\begin{aligned}
\text{ICI}_2 &= (7.74, 7.29, 6.47, 7.58, 7.10) \\
\text{ICI}_3 &= (7, 6.82, 6.91, 7.82, 6.54) \\
\text{ICI}_4 &= (7.33, 6.74, 7.86, 7.43, 6.62) \\
\text{ICI}_5 &= (7.72, 6.67, 7.92, 7.53, 7.31)
\end{aligned}$$

12.2.2.3 Tertiary Assessment Calculation

The value of ICI of the pump manufacturing organization is calculated as follows:

Overall Promotional Factor, $PF = (0.3, 0.3, 0.2, 0.1, 0.1)$

Overall Assessment vector,

$$\text{A} = \begin{bmatrix} 7.50 & 7.04 & 6.52 & 7.18 & 6.86 \\ 7.74 & 7.29 & 6.47 & 7.58 & 7.10 \\ 7.00 & 6.82 & 6.91 & 7.82 & 6.54 \\ 7.33 & 6.74 & 7.86 & 7.43 & 6.62 \\ 7.73 & 6.67 & 7.92 & 7.53 & 7.31 \end{bmatrix}$$

$$\begin{aligned}
ICI &= PF \times A \\
ICI &= (7.48, 7, 6.86, 7.49, 6.89) \qquad (12.3) \\
ICI &= \frac{1}{S}(7.48, 7, 6.86, 7.49, 6.89)
\end{aligned}$$

Where 'S' is the number of subject matter experts evaluated.

$$ICI = 7.14 \epsilon (6, 8)$$

12.2.2.4 Improvement Areas for the Pump Manufacturing Organization

The computation of ICI using MGFA is 7.14. This is in the range (7–8) which shows that the pump organization is likely a digital organization with some effort. This result matched with the atmosphere surfaced at the pump organization. But there is scope for improvement for the pump organization. The areas found for ICI improvements at pump organization are described as follows:

- Proactive risk alerts based on historical data (ICIA_{241})
- Simplify and automate supplier audits (ICIA_{141})
- Investment in technological improvements (ICIA_{313})
- Respond to exceptions before the business is affected (ICIA_{123})

- Enforce health and safety practices ($ICIA_{331}$)
- Reduction of non-value-added costs ($ICIA_{413}$)
- Ensure compliance for procuring timely services ($ICIA_{522}$)

12.2.3 *Proposed Toolkit for the Pump Manufacturing Organization*

In Industry 4.0, the amalgamation of automation and digitization of processes and physical objects will influence human work (Iqbal et al. 2020). With the identified improvement areas based on MGFA, there are specific Industry 4.0 tools and Circularity suggested to the organization. As illustrated in Fig. 12.3, the toolkit is implemented for the identified improvement areas in the pump manufacturing organization.

The vision of the pump manufacturing organization is to deploy tools and systems to enable the supply chain to drive accurate data and eliminate the manual process. The toolkit for integration of Industry 4.0 tools and Circularity is illustrated in Table 12.2.

12.2.4 *Implementation of the Toolkit*

The digital transition to Industry 4.0 starts with an organizational strategy that aligns with Industry 4.0 initiatives (Pozzi et al. 2023). The top management is advised with realistic expectations on the outcome of implementing Industry 4.0 and Circularity.

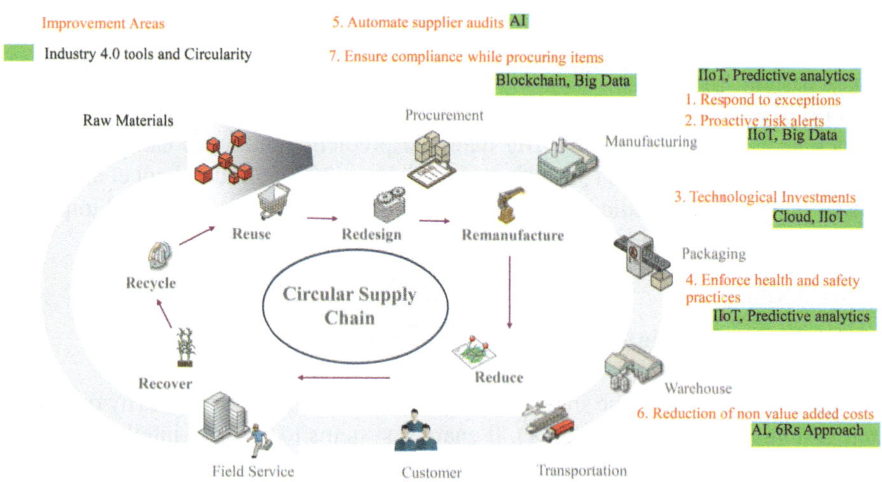

Fig. 12.3 Proposed toolkit for the pump manufacturing organization

Table 12.2 Toolkit for the pump manufacturing organization

Perspective	Identified improvement areas	Industry 4.0 tool/circularity
Technology	Proactive risk alerts based on historical data	Industrial IoT, Big Data Analytics
Business	Simplify and automate supplier audits	Artificial Intelligence
Sustainable development	Investment in technological improvements	Cloud, IoT
Business	Respond to exceptions before the business is affected	Industrial IoT, Predictive Analytics
Sustainable development	Enforce health and safety practices	IoT, Predictive Analytics
Collaboration	Reduction of non-value-added costs	Artificial Intelligence, 6Rs Approach
Management	Ensure compliance for procuring timely services	Blockchain, Big Data Analytics

Following are the specific Industry 4.0 tools and circularity implemented for the pump manufacturing organization.

12.2.4.1 Industrial IoT

Industrial IoT is a technology that integrates sensors, controllers, actuators, and machines with humans allows for the interoperability of organizations, and provides faster access to data (Pivoto et al. 2021). Munirathinam (2020) presented a model using IoT to improve supply chain performance. Based on the analysis, industrial IoT is implemented for the identified improvement areas as illustrated in Table 12.3.

12.2.4.2 Predictive Analytics

Predictive analytics helps the user to predict future outcomes based on historical data (Sarker 2021). It uses data, statistical algorithms, and machine-learning techniques to assess future events. Big data and predictive analytics can be used for supply chain performance (Seyedan and Mafakheri 2020; Sundarakani et al. 2021). Based on the study, Predictive analytics is implemented for the identified improvement areas as illustrated in Table 12.4.

12.2.4.3 Artificial Intelligence

Artificial Intelligence is the simulation of human intelligence processes by machines and systems (Zhang and Lu 2021). It enables systems to perform intellectual tasks such as decision-making, and problem-solving. It has been mentioned that the role of artificial intelligence in the sustainable supply chain for the food and drink

Table 12.3 Industrial IoT

Areas of improvement	Industrial IoT	Technique/Algorithm
Respond to exceptions before the business is affected	Adopt the sensor-driven smart manufacturing to receive a manufacturing quality alert, predict manufacturing cycle time delay, detect anomalies in the equipment	MT Connect adapter should be used to connect to the device. The protocol can be Message Queue Telemetry Transport (MQTT). Data from devices converted into SHDR (Simple Hierarchical Data Representation) format.
Proactive risk alerts based on historical data	Evaluate technology upgrades such as industrial IoT. Transform from preventive to predictive maintenance.	
Investment in technological improvements	The product development time and production efficiency are critical for pump applications. Explore remote monitoring systems such as Radio-frequency identification (RFID), and IoT to examine the components, and assets in real-time, predict future life, and avoid unexpected downtime.	
Enforce health and safety practices	Use IoT devices to monitor service personnel health who is involved in pump maintenance activities in hazardous places or extreme working conditions.	

Table 12.4 Predictive analytics

Areas of improvement	Predictive analytics
Respond to exceptions before the business is affected	Explore predictive maintenance analytics to model foreseeable evolutions of the characteristics of individual assets
Enforce health and safety practices	Leverage sensor-driven internet technology to help field engineers to diagnose an issue on abnormally working high-pressure pumps and get live instructions online to resolve it.

industry (Olan et al. 2021). Based on the analysis, artificial intelligence is implemented for the identified improvement areas as illustrated in Table 12.5.

12.2.4.4 Big Data Analytics

Big data analytics uses advanced analytic techniques on the volume of data that consists of structured, semi-structured, and unstructured data obtained from different sources (Pallamala and Rodrigues 2022). It was identified that big data-driven using a case study approach (Maheshwari et al. 2021). Based on the study, big data analytics is implemented for the identified improvement areas as illustrated in Table 12.6

Table 12.5 Artificial intelligence

Areas of improvement	Artificial intelligence	Technique/Algorithm
Simplify and automate supplier audits	Consolidate supplier details from multiple source/legacy systems into one system to track audit and compliance requirements across suppliers Send automatic notification to all internal stakeholders seeking the feedback of the supplier and add it to master supplier profile automatically. Track supplier risk profile and alert risk mitigation plan across the organization	K-Means Clustering should be used to automate supplier audits. This algorithm allows to clustering the data into different groups such as good and weak suppliers.
Reduction of non-value-added costs	Automate assessment of high-risk suppliers every quarter and evaluate supplier compliance certification periodically. Automate processing supplier change requests with a threshold limit.	

Table 12.6 Big data analytics

Areas of improvement	Big data analytics
Proactive risk alerts based on historical data	Use supplier management big data analytics to analyze the historical patterns to determine price variations and relationship challenges.
Ensure compliance for procuring timely services	Use a virtual contract expert to alert the compliance manager by sending automatic notifications when suppliers do not meet organization compliance requirements. Forecast labor and components necessities. Optimize workload assignment and issue of items to maintenance work orders.

12.2.4.5 Cloud Integration

Cloud integration is a combination of technologies that connects multiple applications, storage systems, and IT infrastructure for the real-time communication of data and processes (Khriji et al. 2022). Cloud computing can be used in supply chain to improve its performance (Novais et al. 2019). Based on the study, cloud integration is implemented for the identified improvement areas as illustrated in Table 12.7.

Table 12.7 Cloud integration

Areas of improvement	Cloud integration
Investment in technological improvements	The product development time and production efficiency are critical for pump applications. Explore remote monitoring systems such as Radio-frequency identification (RFID), IoT to examine the components, assets in real-time, predict future life, and avoid unexpected downtime.

Table 12.8 6Rs approach

Areas of improvement	6Rs approach
Reduction of non-value-added costs	Bring better collaboration between buyer and supplier to refine and update catalogs automatically. Make better replace vs repair analysis.

12.2.4.6 6Rs Approach

Based on the sustainability principle, the waste items can undergo any of the 6Rs of sustainability namely Recover, Reuse, Remanufacture, Recycle, Redesign, and Reduce to make more value out of the items (Liu et al. 2023; Spaltini et al. 2021). 6Rs approach of transition can be used to convert to CSC using renewable energy. Based on the study, the 6Rs approach is implemented for the identified improvement areas as illustrated in Table 12.8.

12.2.4.7 Feasibility of Blockchain

Based on the study, the organization is still needed to have the infrastructure to implement blockchain (Alam et al. 2021; Ronaghi 2021). Blockchain requires a little more maturity and understanding by the stakeholders to implement it.

12.2.5 Final Assessment of the Pump Manufacturing Organization Using MGFA

The developed framework for evaluation is assessed again in the Pump manufacturing organization post-implementation of the toolkit with the subject matter experts. The framework for evaluating the readiness index in SCM for Industry 4.0 is presented in Chap. 2. For the final assessment, 5 experts from five different departments are selected in the respective organization with more than 10 years of professional experience in the supply chain. The experts are asked to evaluate the Industry 4.0 and Circularity factors based on Likert's Scale. Based on the score, the primary assessment is calculated. Similarly, the index for remaining Industry 4.0 and Circularity factors are calculated. Further, the secondary and tertiary assessments are carried out similarly. Based on the computation, Industry 4.0 and Circularity Index (ICI) are derived. The enablers, criteria, and attributes of the supply chain organization are represented by ICIE, ICIC, and ICIA respectively as shown in Table 12.9. A cross-functional team consisting of five Subject Matter Experts (SME) has contributed to rating the factors and providing Promotional Factors (PF).

Table 12.9 Final evaluation of ICI factors for the pump manufacturing industry

ICIEi	ICICij	ICIAijk	SME1	SME2	SME3	SME4	SME5	PFij	PFi	PF
ICIE1	ICIC11	ICIA111	8	9	9	10	8	0.3	0.20	0.30
		ICIA112	8	9	9	8	7	0.3		
		ICIA113	9	8	10	7	9	0.4		
	ICIC12	ICIA121	8	9	10	8	9	0.2	0.40	
		ICIA122	9	8	9	7	9	0.1		
		ICIA123	9	10	8	9	8	0.4		
		ICIA124	10	8	9	8	10	0.3		
	ICIC13	ICIA131	9	10	9	8	9	0.3	0.20	
		ICIA132	8	9	8	7	8	0.4		
		ICIA133	9	9	7	9	9	0.2		
		ICIA134	9	9	10	8	9	0.1		
	ICIC14	ICIA141	8	10	9	9	8	0.4	0.20	
		ICIA142	8	7	9	10	7	0.3		
		ICIA143	9	8	10	8	9	0.3		
ICIE2	ICIC21	ICIA211	9	7	7	9	8	0.4	0.50	0.30
		ICIA212	9	10	8	9	7	0.1		
		ICIA213	8	9	8	10	9	0.3		
		ICIA214	9	8	9	8	7	0.2		
	ICIC22	ICIA221	9	7	8	8	8	0.4	0.20	
		ICIA222	9	10	9	9	7	0.1		
		ICIA223	9	8	9	10	9	0.3		
		ICIA224	7	9	9	9	8	0.2		
	ICIC23	ICIA231	8	9	7	10	7	0.4	0.10	
		ICIA232	9	8	9	8	9	0.5		
		ICIA233	10	9	9	9	6	0.1		
	ICIC24	ICIA241	10	9	9	8	9	0.4	0.20	
		ICIA242	7	9	8	8	9	0.3		
		ICIA243	9	8	10	9	7	0.1		
		ICIA244	9	8	8	10	8	0.2		
ICIE3	ICIC31	ICIA311	9	7	9	9	8	0.4	0.50	0.20
		ICIA312	9	8	7	8	7	0.1		
		ICIA313	9	10	8	9	9	0.3		
		ICIA314	9	8	8	9	9	0.2		
	ICIC32	ICIA321	8	9	9	10	9	0.3	0.20	
		ICIA322	8	9	9	10	8	0.4		
		ICIA323	7	9	8	9	6	0.3		

(continued)

Table 12.9 (continued)

ICIEi	ICICij	ICIAijk	SME1	SME2	SME3	SME4	SME5	PFij	PFi	PF
	ICIC33	ICIA331	9	8	10	9	10	0.50	0.30	
		ICIA332	7	9	7	9	9	0.30		
		ICIA333	8	9	9	9	10	0.20		
ICIE4	ICIC41	ICIA411	7	9	9	7	9	0.30	0.30	0.10
		ICIA412	8	9	10	9	10	0.20		
		ICIA413	9	8	10	8	9	0.50		
	ICIC42	ICIA421	9	10	9	8	8	0.30	0.30	
		ICIA422	6	8	10	9	6	0.20		
		ICIA423	9	8	9	10	8	0.50		
	ICIC43	ICIA431	8	9	8	9	8	0.40	0.20	
		ICIA432	7	9	10	8	9	0.20		
		ICIA433	9	8	9	9	9	0.20		
		ICIA434	9	10	8	7	8	0.20		
	ICIC44	ICIA441	9	8	9	8	10	0.30	0.20	
		ICIA442	7	9	8	10	9	0.50		
		ICIA443	9	8	9	10	6	0.20		
ICIE5	ICIC51	ICIA511	8	9	10	9	8	0.40	0.40	0.10
		ICIA512	9	9	8	10	9	0.20		
		ICIA513	9	9	10	9	10	0.20		
		ICIA514	9	10	8	9	9	0.20		
	ICIC52	ICIA521	9	9	10	8	8	0.40	0.30	
		ICIA522	9	9	10	8	9	0.40		
		ICIA523	10	8	9	8	9	0.20		
	ICIC53	ICIA531	7	9	10	9	8	0.20	0.30	
		ICIA532	9	9	10	8	9	0.30		
		ICIA533	8	9	8	10	9	0.50		

12.2.5.1 Primary Assessment Calculation

The calculation related to the 'Services Management' criterion is presented as follows:

Promotional Factor of 'Services Management' criterion, $PF_{11} = (0.3, 0.3, 0.4)$

Evaluation vector related to 'Services Management' criterion is represented by

$$A_{11} = \begin{bmatrix} 8 & 9 & 9 & 10 & 8 \\ 8 & 9 & 9 & 8 & 7 \\ 9 & 8 & 10 & 7 & 9 \end{bmatrix}$$

The ICI of an Organization is the product of overall Assessment Factor

A and Promotional Factor PF.

ICI pertaining to 'Services Management' criterion is represented by

$$ICI_{11} = PF11 \times A_{11}$$
$$ICI_{11} = (8.40, 8.60, 9.40, 8.20, 8.10)$$

Similarly, the ICI for the remaining criteria is calculated as follows:

$$ICI_{12} = (9.10, 9.00, 8.80, 8.30, 8.90)$$
$$ICI_{13} = (8.60, 9.30, 8.30, 7.80, 8.60)$$
$$ICI_{14} = (8.30, 8.50, 9.30, 9.00, 8.00)$$
$$ICI_{21} = (8.70, 8.10, 7.80, 9.10, 8.00)$$
$$ICI_{22} = (8.60, 8.00, 8.60, 8.90, 8.20)$$
$$ICI_{23} = (8.70, 8.50, 8.20, 8.90, 7.90)$$
$$ICI_{24} = (8.80, 8.70, 8.60, 8.50, 8.60)$$
$$ICI_{31} = (9.0, 0, 8.20, 8.30, 8.90, 8.40)$$
$$ICI_{32} = (7.70, 9.00, 8.70, 9.70, 7.70)$$
$$ICI_{33} = (8.20, 8.50, 8.90, 9.00, 9.70)$$
$$ICI_{41} = (8.20, 8.50, 9.70, 7.90, 9.20)$$
$$ICI_{42} = (8.40, 8.60, 9.20, 9.20, 7.60)$$
$$ICI_{43} = (8.20, 9.00, 8.60, 8.40, 8.40)$$
$$ICI_{44} = (8.00, 8.50, 8.50, 9.40, 8.70)$$
$$ICI_{51} = (8.60, 9.20, 9.20, 9.20, 8.80)$$
$$ICI_{52} = (9.20, 8.80, 9.80, 8.00, 8.60)$$
$$ICI_{53} = (8.10, 9.00, 9.00, 9.20, 8.80)$$

12.2.5.2 Secondary Assessment Calculation

The calculation related to the 'Business Perspective' enabler is presented as follows:

Promotional Factor of 'Business Perspective' enabler, $PF_1 = (0.2, 0.4, 0.2, 0.2)$

Evaluation vector related to 'Business Perspective' enabler is represented by

$$A_1 = \begin{bmatrix} 8.40 & 8.60 & 9.40 & 8.20 & 8.10 \\ 9.10 & 9.00 & 8.80 & 8.30 & 8.90 \\ 8.60 & 9.30 & 8.30 & 7.80 & 8.60 \\ 8.30 & 8.50 & 9.30 & 9.00 & 8.00 \end{bmatrix}$$

ICI of 'Business Perspective' enabler is represented by

$$\mathrm{ICI}_1 = \mathrm{PF1} \times \mathrm{A1}$$
$$\mathrm{ICI}_1 = (8.70,\ 8.88,\ 8.92,\ 8.32,\ 8.50)$$

Similarly, the ICI for remaining enablers is calculated as follows:

$$\mathrm{ICI}_2 = (8.70,\ 8.24,\ 8.16,\ 8.92,\ 8.15)$$
$$\mathrm{ICI}_3 = (8.50,\ 8.45,\ 8.56,\ 9.09,\ 8.65)$$
$$\mathrm{ICI}_4 = (8.22,\ 8.63,\ 9.09,\ 8.69,\ 8.46)$$
$$\mathrm{ICI}_5 = (8.63,\ 9.02,\ 9.32,\ 8.84,\ 8.74)$$

12.2.5.3 Tertiary Assessment Calculation

The value of ICI of the Pump manufacturing organization is calculated as follows:

Overall Promotional Factor, PF = (0.3, 0.3, 0.2, 0.1, 0.1)

Overall Assessment vector,

$$\mathrm{A} = \begin{bmatrix} 8.70 & 8.88 & 8.92 & 8.32 & 8.50 \\ 8.70 & 8.24 & 8.16 & 8.92 & 8.15 \\ 8.50 & 8.45 & 8.56 & 9.09 & 8.65 \\ 8.22 & 8.63 & 9.09 & 8.69 & 8.46 \\ 8.63 & 9.02 & 9.32 & 8.84 & 8.74 \end{bmatrix}$$

$$\mathrm{ICI} = \mathrm{PF} \times \mathrm{A}$$
$$\mathrm{ICI} = (8.6,\ 8.59,\ 8.68,\ 8.74,\ 8.45)$$
$$\mathrm{ICI} = \frac{1}{S}(8.6,\ 8.59,\ 8.68,\ 8.74,\ 8.45)$$

where 'S' is the number of subject matter experts evaluated.

$$ICI = 8.61 \epsilon (8, 10)$$

Final ICI for pump manufacturing organization = 8.61

Based on the assessment of Industry 4.0 and Circularity factors, the ICI derived using MGFA for the pump manufacturing organization is 8.61. This is in the range of (8–10) which shows that the pump manufacturing organization is exceeding on digital and circularity transformation. The results have coincided with the practical environment of the organization.

12.3 Results and Discussions

The ICI for case organizations before and after implementation of Industry 4.0 tools and Circularity practices suggests that there are significant improvements in the areas identified as illustrated in Fig. 12.4. The statistical validation carried out before and after implementation revealed that the results are practical and sustainable. Increased visibility and agility provided by Industry 4.0 tools and Circularity practices have a considerable role in providing flexible, agile, robust, and sustainable supply chains.

12.4 Structure of the Roadmap

The first phase in the roadmap to Industry 4.0 and Circularity transformation is the 'Analysis' phase, as illustrated in Fig. 12.5. In this phase, the key managerial requirements are analyzed to know the knowledge level of the stakeholders and it is important to create awareness on Industry 4.0 and Circular practices by conducting a series of workshops. Further, the opportunities and risks are analyzed based on the business cases provided. The objective of this phase is to maximize the strengths and minimize the risks of the organization. The project scope is defined, and the project team is formulated. In general, the timeline for this phase is 1–3 months.

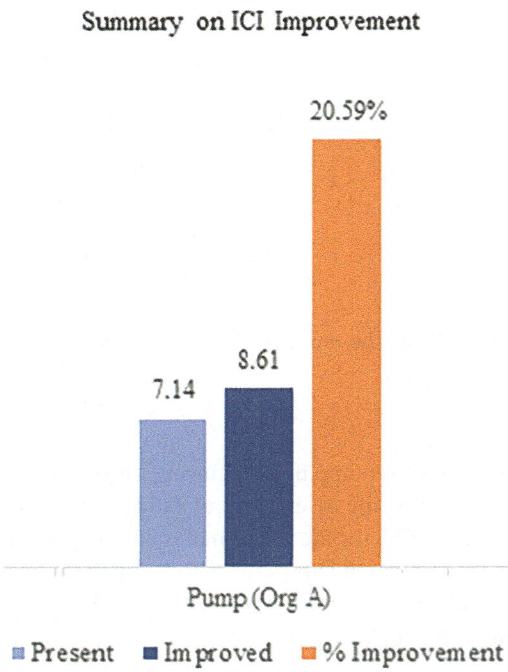

Fig. 12.4 Before and after implementing Industry 4.0 and circularity

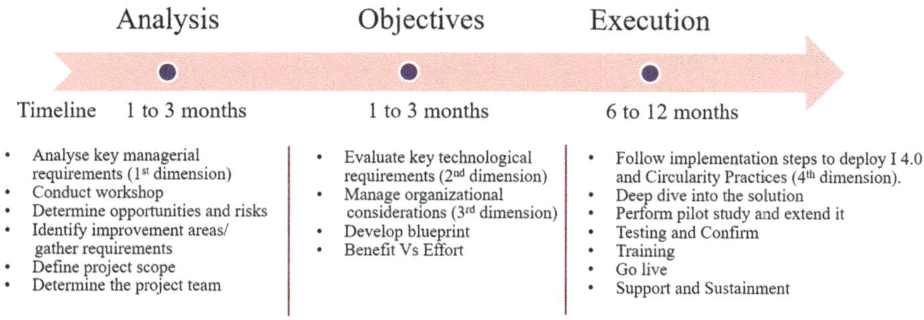

Fig. 12.5 Structure of roadmap to Industry 4.0 and circularity transformation

The second phase is 'Objectives setting' which is to evaluate key technological requirements and manage organizational considerations based on the project goals. In this phase, the existing competencies and the current business process are studied. For the identified improvement areas and requirements, a blueprint is developed. The benefit versus effort analysis is made and aligned with the stakeholders. In general, the timeline for the 'Objective setting' phase is 1–3 months.

The third phase is the 'Execution' phase where the real implementation of the industry 4.0 and Circular practices is conducted. In this phase, a detailed solution is developed, and a pilot study is performed. Based on the results, the solution is extended to other requirements. User acceptance testing is performed and confirmed. Later, the training is organized with the workforce and the system goes live. Further, hyper-care support and sustainment are provided till the system is stabilized. It is advisable to start with a smaller unit and expand it to larger units so that the initial problems are identified and addressed at the early stage.

The specific goals of the three organizations are listed in Table 12.10. The goals of the pump manufacturing organization are to improve transparency, guided buying, and improve IT infrastructure. The goals of paper manufacturing organizations are to predict raw material shortage, remote monitoring of resources, and reduce production wastage. The goals of the construction industry are to manage spending, preventive maintenance, real-time information sharing, energy monitoring, and management. The specific goals of the respective organization are linked to the common goal. The common goal is to transform into a digital organization for better connectivity, monitoring, and control capabilities and follow circular practices to minimize environmental impacts and improve resource efficiency. Successful long-term implementation of the developed roadmap can be reached by setting the organizational goals precisely. The total timeline to implement the roadmap will take 12–15 months based on the complexity of the supply chain processes. Based on the experience gained from three case organizations, the generalized roadmap is proposed. The generalized roadmap consists of three phases namely the analysis phase, objectives setting phase, and implementation phase. The four dimensions namely key managerial requirements, key technological requirements, key organization requirements, key implementation steps are part of the three phases of the roadmap.

Table 12.10 Consolidated roadmap to Industry 4.0 and circularity transformation

Case organization	Pump			
Type	Discrete Manufacturing			
Specific Goals	Improve transparency, guided buying, improve IT infrastructure			
Common Goal	Transform into a digital organization for better connectivity, monitoring, and control capabilities and follow circular practices to minimize environmental impacts and improve resource efficiency.			
Phases with timeline	Analyse (1–3 months) → Objectives Setting (1–3 months) → Execution (6–12 months)			
Total timeline for implementation	12–15 months (based on the complexity of supply chain process)			
Toolkits	Industrial IoT, AI, Cloud, Big Data Analytics, 6Rs approach			
Dimensions	Key managerial requirements	Key technological requirements	Key organizational requirements	Key implementation steps
Stages	1-Connect	2-Commence	3-ConFigureure	4-Confirm
	Define scope	Develop blueprint	Integrate systems	Go live and sustainment
Extended scope	Digital connectivity with suppliers		Digital connectivity with customers	

The scope can be extended to suppliers and customers provided the supply chain members are also interested to move towards digital and circularity transformation.

12.5 Conclusion

In summary, a meticulous review of pump manufacturing with Industry 4.0 tools, and Circularity practices has unfolded through an exhaustive literature review. With the insights from industry experts and stakeholders, the key factors shaping an organization's readiness are studied for the evolution into Industry 4.0 and Circularity. This review has laid the foundation for developing a robust framework, designed to assess organizational preparedness in these transformative areas. With the maturity of supply chain processes and a demonstrated eagerness to embrace digital technologies and circular strategies, a case organization from the pump industry was strategically chosen. Employing the developed framework, our research employs the Multi-Grade Fuzzy Approach (MGFA) to scrutinize the current state of readiness for Industry 4.0 and Circularity within each selected organization. The outcomes of this assessment spotlight specific improvement areas unique to each case organization, offering a roadmap for targeted enhancements.

To validate the practical applicability of Industry 4.0 tools and circularity practices, a rigorous statistical validation is executed. The culmination of these efforts materializes in the formulation and subsequent implementation of a specialized toolkit for the chosen case organization, seamlessly aligning with the tenets of Industry 4.0 and Circularity. This toolkit undergoes meticulous analysis and refinement in collaboration with subject matter experts, ensuring its relevance and effectiveness. The concluding phase of this research entails a final assessment of the factors influencing Industry 4.0 and Circularity transformation within the pump industry. Validation, achieved through interactive sessions with subject matter experts, strengthens the findings. A comprehensive roadmap for Industry 4.0 and Circularity transformation is provided. This roadmap not only stands as a testament to the research's outcomes but also serves as a pragmatic guide for organizations across industries, steering them through the transformative journey towards Industry 4.0 and Circularity.

References

Afeltra G, Alerasoul SA, Strozzi F (2023) The evolution of sustainable innovation: from the past to the future. Eur J Innov Manag 26(2):386–421. https://doi.org/10.1108/EJIM-02-2021-0113

Alam S, Shuaib M, Khan WZ, Garg S, Kaddoum G, Hossain MS, Zikria YB (2021) Blockchain-based initiatives: current state and challenges. Comput Netw 198:108395. https://doi.org/10.1016/j.comnet.2021.108395

Fosci P, Psaila G (2024) A unified view of multi-grade fuzzy-set models in J-CO-QL+. Neurocomputing 565:126968. https://doi.org/10.1016/j.neucom.2023.126968

Iqbal A, Zhao G, Suhaimi H, He N, Hussain G, Zhao W (2020) Readiness of subtractive and additive manufacturing and their sustainable amalgamation from the perspective of Industry 4.0: a comprehensive review. Int J Adv Manuf Technol 111:2475–2498. https://doi.org/10.1007/s00170-020-06287-6

Khriji S, Benbelgacem Y, Chéour R, Houssaini DE, Kanoun O (2022) Design and implementation of a cloud-based event-driven architecture for real-time data processing in wireless sensor networks. J Supercomput 78(3):3374–3401. https://doi.org/10.1007/s11227-021-03955-6

Liu AR, Lin ZJ, Wei M, Tang Y, Zhang H, Peng XG et al (2023) The potent analgesia of intrathecal 2R, 6R-HNK via TRPA1 inhibition in LF-PENS-induced chronic primary pain model. J Headache Pain 24(1):141. https://doi.org/10.1186/s10194-023-01667-1

Maheshwari S, Gautam P, Jaggi CK (2021) Role of Big Data analytics in supply chain management: current trends and future perspectives. Int J Prod Res 59(6):1875–1900. https://doi.org/10.1080/00207543.2020.1793011

Munirathinam S (2020) Industry 4.0: industrial internet of things (IIOT). In: Advances in computers, Vol. 117, No. 1. Elsevier, Cambridge, MA, pp 129–164. https://doi.org/10.1016/bs.adcom.2019.10.010

Novais L, Maqueira JM, Ortiz-Bas Á (2019) A systematic literature review of cloud computing use in supply chain integration. Comput Ind Eng 129:296–314. https://doi.org/10.1016/j.cie.2019.01.056

Olan F, Suklan J, Arakpogun EO, Robson A (2021) Advancing consumer behavior: the role of artificial intelligence technologies and knowledge sharing. IEEE Trans Eng Manag 71:13227–13239. https://doi.org/10.1109/TEM.2021.3083536

Pallamala RK, Rodrigues P (2022) An investigative testing of structured and unstructured data formats in big data application using apache spark. Wirel Pers Commun 122(1):603–620. https://doi.org/10.1007/s11277-021-08915-0

Pivoto DG, De Almeida LF, da Rosa Righi R, Rodrigues JJ, Lugli AB, Alberti AM (2021) Cyber-physical systems architectures for industrial internet of things applications in Industry 4.0: A literature review. J Manuf Syst 58:176–192. https://doi.org/10.1016/j.jmsy.2020.11.017

Pozzi R, Rossi T, Secchi R (2023) Industry 4.0 technologies: critical success factors for implementation and improvements in manufacturing companies. Prod Plan Control 34(2):139–158. https://doi.org/10.1080/09537287.2021.1891481

Ronaghi MH (2021) A blockchain maturity model in agricultural supply chain. Inf Process Agri 8(3):398–408. https://doi.org/10.1016/j.inpa.2020.10.004

Sarker IH (2021) Data science and analytics: an overview from data-driven smart computing, decision-making and applications perspective. SN Comp Sci 2(5):377. https://doi.org/10.1007/s42979-021-00765-8

Seyedan M, Mafakheri F (2020) Predictive big data analytics for supply chain demand forecasting: methods, applications, and research opportunities. J Big Data 7(1):53. https://doi.org/10.1186/s40537-020-00329-2

Spaltini M, Poletti A, Acerbi F, Taisch M (2021) A quantitative framework for Industry 4.0 enabled Circular Economy. Procedia CIRP 98:115–120. https://doi.org/10.1016/j.procir.2021.01.015

Sundarakani B, Pereira V, Ishizaka A (2021) Robust facility location decisions for resilient sustainable supply chain performance in the face of disruptions. Int J Logist Manag 32(2):357–385. https://doi.org/10.1108/IJLM-12-2019-0333

Vimal KEK, Kulatunga AK, Ravichandran M, Kandasamy J (2021) Application of multi grade fuzzy approach to compute the circularity index of manufacturing organizations. Procedia CIRP 98:476–481. https://doi.org/10.1016/j.procir.2021.01.137

Zhang C, Lu Y (2021) Study on artificial intelligence: the state of the art and future prospects. J Ind Inf Integr 23:100224. https://doi.org/10.1016/j.jii.2021.100224

Index

J. Kandasamy et al., *Circular Supply Chain Management*, Circular Economy and Sustainability, https://doi.org/10.1007/978-3-031-81750-2

The manufacturer's authorised representative in the EU is Springer Nature Customer Service Centre GmbH, Europaplatz 3, 69115 Heidelberg, Germany. If you have any concerns regarding our products, please contact ProductSafety@springernature.com

Printed and bound by CPI Group (UK) Ltd, Croydon, CR0 4YY
30/09/2026
02225303-0010